Number 4 1995

Wind Blown Sediments in the Quaternary Record

Series Editor
John Lowe

Centre for Quaternary Research,
Department of Geography,
Royal Holloway,
University of London,
Egham,
Surrey TW20 0EX, UK.

Edited by
Edward Derbyshire

Published on behalf of the
Quaternary Research Association
by John Wiley & Sons

Copyright 1995 by <u>Quaternary Research Association</u>, Cambridge, UK.

Published 1995 by   John Wiley & Sons Ltd,
                        Baffins Lane, Chichester,
                        West Sussex PO19 1UD, England

                        National   01243 779777
                        International   (+44) 1243 779777

Other Wiley Editorial Offices

John Wiley & Sons, Inc., 605 Third Avenue,
New York, NY 10158-0012, USA

Jacaranda Wiley Ltd, 33 Park Road, Milton,
Queensland 4064, Australia

John Wiley & Sons (Canada) Ltd, 22 Worcester Road,
Rexdale, Ontario M9W 1L1, Canada

John Wiley & Sons (SEA) Pte Ltd, 37 Jalan Pemimpin #05-04,
Block B, Union Industrial Building, Singapore 2057

British Library Cataloguing in Publication Data

A catalogue record for this book is available from the British Library

Recommended referencing: Professor Edward Derbyshire (ed.). 1995. *Wind Blown Sediments in the Quaternary Record*, Quaternary Proceedings No. 4, John Wiley & Sons, Chichester, UK.

ISBN 0-471-95860-3

Camera-ready copy prepared by City Cartographic, London Guildhall University.
Printed and bound in Great Britain by Bookcraft (Bath) Ltd.

CONTENTS

Professor Wang Jingtai
1935-1994

This volume is dedicated to the memory of Wang Jingtai, exceptional Quaternary scientist, valued colleague, sensitive and loyal friend who was born, lived and died on the loess.

FOREWORD

A new collaboration between *John Wiley & Sons* and the *Quaternary Research Association.*

The publication of this fourth issue of the *Quaternary Proceedings* series marks the launch of a new collaboration between *John Wiley & Sons* and the *QRA*. The QP series was originally produced entirely 'in-house' by the *QRA*, but it has rapidly out-stripped the association's capacity to handle the demands of international sales and promotion. In seeking a partner to assist in the process, the association readily accepted a contract with *John Wiley & Sons*. The two organisations already collaborate over the production of *Journal of Quaternary Science*, the business of which has always been conducted to very high professional standards and in a warm spirit of co-operation. We look forward to many more years of such co-operation in the future.

Quaternary Proceedings publishes the proceedings of scientific conferences and other meetings held within the aegis of the *QRA* or organised jointly by the *QRA* (or its members) and other scientific organisations. The aim of the series is to enable the rapid publication of important scientific information, especially that which addresses topical themes of interest to the international community of Quaternary scientists.

Contributions to the series are all peer-reviewed (minimum two referees). The contents of each issue and the refereeing of contributions is handled by the editors appointed to each issue. Editing and production matters are 'supervised' by the *Series Editor*, John Lowe. Each issue is typeset by the *City Cartographic*

Unit of London Guildhall University under the direction of Don Shewan, using an Apple Macintosh system and output on a high resolution Imagesetter. Wherever possible, contributions should be supplied as both hard copy and on diskette, using any of the common word processing formats. Printing, promotion and sales are the responsibility of *John Wiley & Sons*, under the direction of Helen Bailey. Proposals for new volumes should be directed to the *Series Editor*. Enquiries concerning sales should be directed to *John Wiley & Sons*.

We thank the following who have assisted in the production of this volume: Peter Coxon, Wishart Mitchell, Frank Oldfield and Mike Walker (*QRA* Executive Committee) who were heavily involved in the initial contract agreements; Abi Hudlass, Louise Portsmouth and Claire Walker (*Wiley*) who handled production matters.

Helen Bailey
Editor, Environmental and Earth Sciences
John Wiley & Sons, Chichester

J. John Lowe
Director, Centre for Quaternary Research
Royal Holloway, University of London

Don Shewan
Director, City Cartographic Unit
London Guildhall University

February, 1995

PREFACE

This volume consists of a selection of papers presented at a joint meeting of the Quaternary Research Association (QRA) and the Commissions on Stratigraphy, Glacial Deposits, Loess, and Palaeopedology of the International Union for Quaternary Research (INQUA) held at Royal Holloway, University of London 4-7 January 1994.

As the organiser of the meeting, I am aware that without the encouragement and support of a number of colleagues and institutions, the meeting would not have taken place. First, I must thank the Committee of the QRA for their early and warm encouragement. Equally, I received strong support from my fellow INQUA Executives, especially President Liu Tungsheng and Treasurer Eduardo de Mulder. Vital to the realisation of the meeting was the enthusiasm of senior staff in the Department of Geography, Royal Holloway, University of London, especially Professor Jim Rose. Not surprisingly, the office and cartographic staff played a key role as the conference date approached, and I am pleased to record my sincere thanks to Pam Cardwell, Kathy Roberts, June Brain and Justin Jacyno who ensured that we had an Abstract Volume in good time.

Once started, the conference would have faltered more than it did without the energy and unselfish enthusiasm of Royal Holloway's postgraduate students in physical geography. My sincere thanks go to Fiona Clayton, Sarah Dixon, Meng Xingmin, John Newstead, Natalie Perkins, Shaun Richardson, Gingy Robson, Mahdo Sah, Jean-Luc Schwenninger, Milap Sharma, Charlie Sheldrick, Wang Jianmin and Phil Wood.

The production of this volume owes much to the considerable editorial assistance and scientific judgement of more than twenty academic referees. In particular, I must acknowledge the encouragement, tolerance and attention to detail of Professor John Lowe as Senior Editor, and the efficiency and professionalism of Don Shewan at the production stage.

Edward Derbyshire
Centre for Quaternary Research
Royal Holloway, University of London

February 1995

INTRODUCTION

Investigation of the atmosphere-ocean system has been recognised for many years as a key route to an improved understanding of the global climate, and the deep ocean sedimentary record has proved a keystone in this structure. The complexity of the system has been highlighted by realisation of the wide-ranging effects upon the system of events that are rapid and regional in scale. Records from the land hemisphere generally proved to be less complete and of lower resolution than those from the oceans until the potential of the ice cores was realised. Study of the great loess successions of central and eastern Asia has evolved at a different, but accelerating pace: classic lithostratigraphy 1970-1985; high resolution magnetostratigraphy and magnetic susceptibility climate-proxy work 1985-1992; and magnetic mineralogy and very high resolution grain size climate proxy work, with the initial estimates of palaeoprecipitation for North China using mineral magnetics, in the period 1992-1995. Improvements in methods of dating aeolian silts coincided with the demonstration of the potential sensitivity of the central Asian loess as an index of palaeoclimate: work began in 1993 on loess sections which appear to offer a resolution certainly of century quality and, in at least one case, of decadel quality. To the mineral magnetics approach to palaeoclimate has been added the subtle and sensitive soil micromorphology method. These extremely high resolution land records have been shown, thus far only in outline, to be highly responsive to the dynamics of the East Asian monsoon. Monsoonal dynamics, in turn, have been shown to be related, perhaps in stepwise form (Liu, T.S. & Ding, Z.L. 1993 *Global and Planetary Change* 7: 119-130), to the rate and periodicity of uplift of the Tibetan Plateau. In GCM work, it is generally agreed that the Tibetan Plateau is a key component, together with northern hemisphere ice volumes.

This volume contains nine papers that illustrate several aspects of the aeolian Quaternary record as a general theme. There is a strong emphasis on the impressively thick stratigraphical records provided by the loess-palaeosol sequences of China and central Asia. The papers examine data derived from a range of complementary techniques: the marine Oxygen isotope record, magnetostratigraphy, spectral analysis, magnetic susceptibility, anisotropy of magnetic susceptibility, rock magnetism, magnetic grain size, magnetic separation using the citrate-bicarbonate-dithionite technique, particle size analysis, palaeosol micromorphology, scanning electron microscopy, carbon isotopes, electron spin resonance, and luminescence dating.

The first paper considers a detailed magnetic susceptibility record from Karamaidan in Tadjikistan, and demonstrates that a clear similarity exists between that region, several Chinese loess sections, and the marine Oxygen isotope record. **Shackleton** *et al.* show that dust accumulation in China and Tadjikistan is more highly coherent with global ice volume as expressed by the marine isotope record than with either the orbital insolation curve or an orbitally-modelled ice volume record. This paper is followed by a consideration by **Derbyshire** *et al.* of some practical problems encountered in correlating loess-palaeosol sequences. They show that some sections used for climate-proxy work, although of high resolution, are incomplete. Other problems, including massive lowering of a loess plateau surface by human action over 1000 years ago and 'welding' of one palaeosol on another, have important implications for regional stratigraphic correlation especially when combined with the common stratigraphic practice of 'counting down from the top'. Critical evaluation of the suitability of loessic sediments as climate proxies is also provided by **Clarke** who considers a number of magnetic fabrics from loess exposures straddling the Tibetan Front. Aeolian (air fall), fluvially-reworked and slope-reworked fabrics are discriminated.

Rock magnetic measurements of a high resolution loess-palaeosol sequence near Xining, north-eastern Tibet, are shown by **Chen** *et al.* to match closely other stratigraphic data. By combining data on magnetic property and magnetic grain size, they signal a means of separating the effects of *in situ* processes from those controlling transport of wind-blown material. Evidence of high frequency variations in the degree of 'magnetic pedogenesis' also raise the possibility of producing a record comparable to that provided by the ice cores.

The source of the loess making up the great accumulations in North China has been debated for decades. In a second contribution, **Clarke** brings to bear on this question the evidence provided by rare earth elements (REE). She is able to show that the loessic silts of Lanzhou have a remarkably similar signature to silts in the mountains of northern Tibet, suggesting that the Tibetan Plateau was a significant source of material for the western margins of the Loess Plateau.

The paper by **Liu** *et al.* sheds further light on the contentious topic of the relative amount of ferrimagnetic material found in the loess and palaeosol series of North China. They show that the suggestion that the citrate-bicarbonate-dithionite technique can separate the pedogenic and primary ferrimagnetic components in loess is an oversimplification both of technique and of the spatial variation in the magnetic grain size distribution.

The last three papers focus on palaeosols and pedogenesis. In a wide-ranging review, **Catt** considers that the potential of buried and surface palaeosols for palaeoclimate interpretation is restricted at present by a certain poverty of knowledge about the complex relationships between soil properties and climatic factors, and by the recurrent problem of estimating the length of soil-forming intervals. A call is made for a new definition of 'palaeosol' and for a new system by which to classify palaeosols. The future is seen as one in which multivariate models of the soil-landscape continuum, using geographical information systems, will be used to overlay comprehensive data sets of temporal and spatial variations in soils.

The paper by **Bronger** *et al.* brings the reader back to Tadjikistan, this time for a consideration of the climatic data provided by the palaeosols within the thick loess sequences, especially at Karamaidan and Chashmanigar. Using micromorphology and other techniques, the authors provide clear evidence of clay illuviation, and show that all strongly developed B or Bt soil horizons represent interglacials similar to the Holocene. A close match is demonstrated between the loess-palaeosol sequence at Karamaidan and the SPECMAP curve for the Brunhes Chron. It is suggested that this region provides the most detailed land record of the Matuyama Chron in central Asia, and that the palaeoclimatic information it contains is superior to that found in either the North China loess or the deep sea record.

The final contribution shows that, given their closeness to the Mediterranean coast, the sedimentary fills within the Mount Carmel caves of Israel contain an important aeolian constituent. The sedimentology and modification by weathering of such old (Lower Palaeolithic) deposits have been little studied. Data presented here show that these materials have suffered strong post-depositional re-working involving anthropogenic and biological processes, groundwater effects, and chemical weathering. Although differentiation between penecontemporaneous and later modification of the cave deposits remains problematical, **Tsatskin** *et al.* suggest that, on the basis of micromorphology, this modification has included pedogenesis.

This collection of papers represents less than one fifth of the scientific contributions debated at the conference. The reader might like to note that a companion volume entitled "Aeolian Sediments and the Quaternary Record" will appear as a special issue of *Quaternary Science Reviews* before the end of 1995, and that the conference Abstract Volume is still available.

Edward Derbyshire
Centre for Quaternary Research
Royal Holloway, University of London.

February 1995

Quaternary Proceedings No. 4, 1995 1-6

Accumulation Rate of Loess in Tadjikistan and China: Relationship with Global Ice Volume Cycles.

N.J. Shackleton, Z. An, A.E. Dodonov, J. Gavin, G.J. Kukla, V.A. Ranov and L.P. Zhou.

Shackleton, N.J., An, Z., Dodonov, A.E., Gavin, J., Kukla, G.J., Ranov, V.A. and Zhou, L.P., 1995 Accumulation rate of loess in Tadjikistan and China: relationship with global ice volume cycles, In *Wind Blown Sediments in the Quaternary Record* (Edward Derbyshire). Quaternary Proceedings No. 4, John Wiley & Sons Ltd., Chichester, pp. 1- 6.

Abstract

Thick loess sequences provide valuable long-term records of dust accumulation in northern China and in former Soviet Central Asia. We report a detailed record of magnetic susceptibility for the Karamaidan loess section in Tadjikistan, for the Brunhes magnetochron. We have correlated this record with those obtained from three Chinese loess sections and with the marine oxygen isotope record. The patterns exhibited by these records over the Middle Pleistocene are remarkably similar. Using a modified version of the susceptibility age model of Kukla et al. (1988), we obtained records of varying dust accumulation in China and Tadjikistan for the last 2.6 and 0.8 million years respectively. Cross-spectral analysis shows that the dust accumulation in both China and Tadjikistan is more highly coherent with global ice volume as represented by the marine oxygen isotope record than with orbital insolation or with an orbitally modelled ice volume record. This suggests that the accumulation of dust is causally related to the development of continental ice sheets rather than being directly controlled by insolation variations.

KEYWORDS: magnetic susceptibility; dust records; cross-spectral analysis.

Shackleton, N.J. and Zhou, L.P., Godwin Laboratory for Quaternary Research, University of Cambridge, Free School Lane, Cambridge CB2 3RS, UK.

An, Z., Xi'an Laboratory of Loess & Quaternary Geology, Chinese Academy of Sciences, PO Box 17, Xi'an, China.

Dodonov, A.E., Geological Institute, Russian Academy of Sciences Pyzhevsky per. 7, 109017 Moscow, Russia, CIS.

Gavin, J. and Kukla, G.J., Lamont-Doherty Earth Observatory, Palisades, NY 10964, USA.

Ranov, V.A., Institute of History, Archaeology and Ethnography, 33 Rudaki Ave., Dushanbe, 737025 Republic of Tadjikistan, CIS.

Zhou, L.P., The McDonald Institute for Archaeological Research, University of Cambridge, Downing Street, Cambridge CB2 3ER, UK.

Introduction

While many continuous records of Quaternary climate are available from ocean sediments (CLIMAP project members, 1976, 1981), most of the few long, continuous records of past climate from land sites are from those areas covered by loess. The most extensively studied region is in the Chinese Loess Plateau (Liu 1988, Liu 1991), but thick loess sequences also exist in Tadjikistan where the presence of numerous Palaeolithic stone tools in the interbedded soils provides additional interest (Dodonov 1991). In the loess sequences of both areas, as well as in the less extensive and more discontinuous loess sections of central Europe (Kukla 1977), the most obvious evidence of changing climate is buried soils formed during relatively mild and moist periods separating intervals of dry and cool glacial climate. As magnetic susceptibility is found to be low in the loess and high in the intervening soils, it provides a convenient palaeoclimate proxy (Heller & Liu 1982, Kukla et al. 1988).

Magnetic susceptibility data from three thick Chinese loess sections (Xifeng 1, Xifeng II and Luochuan) have already been published (Kukla et al. 1990). Here we report the first long and detailed susceptibility record from the Karamaidan section in Tadjikistan (38°38'N, 68°51'E), shown in Figure 1 with the upper part of the Xifeng II section for comparison. Susceptibility is high in the soil horizons and low in the intervening loesses. Most of the visible fossil soils at this locality are polygenetic and hence referred to as pedocomplexes (PC) I, PCII and so on. Note that there has been more than one scheme for numbering the soils. The numbering scheme used here is as used in Karamaidan by Heller (personal comm.) and by Bronger et al. (this volume). It is consistent with the scheme of Lazarenko et al. (1981), but differs from that used in Dodonov (1991). First

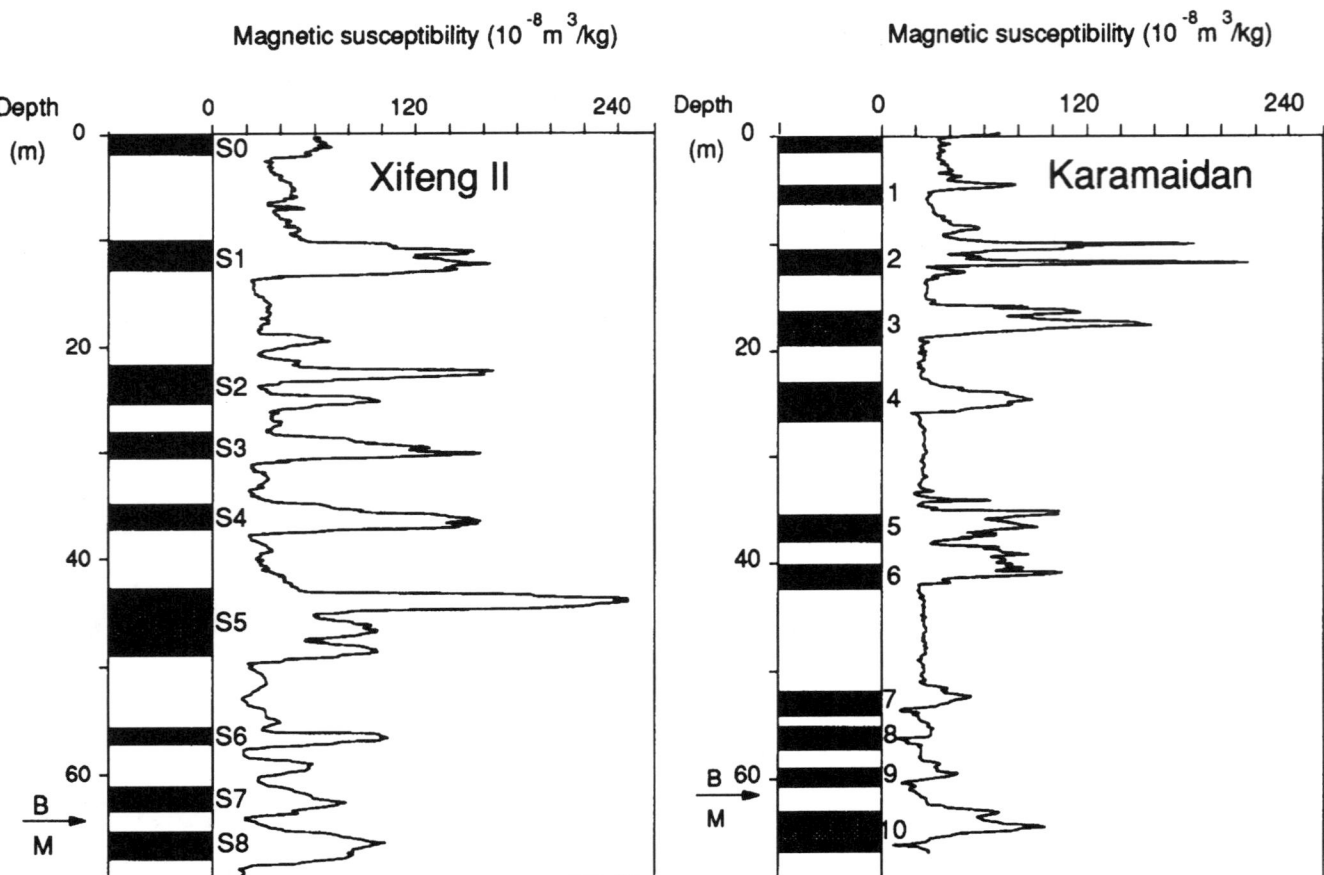

Figure 1. Magnetic susceptibility (10^{-8} m³/kg) versus depth in Xifeng II (upper 70 m only) and Karamaidan. Soils are shown in black and labelled from the top down. The Brunhes/Matuyama boundary is marked by an arrow. More than one system has been used for the numbering of the pedocomplexes in Tadjikistan so that the same horizon may be referenced differently by individual researchers. The numbering used here is as used in Karamaidan by F. Heller (personal comm.) and by A. Bronger (personal comm.) and agrees with Lazarenko et al.'s 1981 scheme but differs from that used in Dodonov (1991).

order age control is established by the position of the Brunhes/Matuyama (B/M) boundary (Pen'kov 1982).

Published interpretations of the loess stratigraphy in Tadjikistan are quite different from those in China (Dodonov 1991; Lazarenko et al. 1981). In China, the mean sedimentation rate of the loess/soil cycles has been fairly constant over the past million years, but in Tadjikistan it has been supposed that most of the deposits accumulated within the last 300 ka and that the accumulation before that time was much lower, either PC3 or PC4 being correlated with the last interglacial. This interpretation was based on early thermoluminescence (TL) measurements that are no longer regarded as yielding meaningful ages (Wintle & Huntley 1982). Our current TL dating study suggests that TL data below PC1 provide only minimum age estimates, and that PC3 is definitely older than the last interglacial. The published chronology has confused interpretation of the Palaeolithic archaeology of Central Asia. Karatau culture artifacts have been found in PC4 and PC5 in about 30 sites: this pebble tool industry is characterized by archaic forms (Ranov & Davis 1979; Ranov 1988) which elsewhere in Asia are considered to be much older than the assigned age of approximately 100 ka. A new chronology for the Tadjikistan loess sequence is needed to resolve these geological and archaeological controversies.

Age Model Development

Kukla et al. (1988) proposed a method of dating long loess

sequences in China. This is based on the observation that unweathered loess has very low magnetic susceptibility and may be regarded as a varying dilutant of the magnetic component. To a first order approximation, the flux (formation and/or deposition) of the susceptibility carrier is considered constant and the magnetic concentration in the sediment is taken as a measure of dilution by the nonmagnetic dust. The age of any layer can then be determined by calibrating the accumulation of the magnetic component between consecutive boundaries of known age. Strict application of this model requires all sedimentation to be subaerial, with no secondary erosion. Such conditions are usually fulfilled in subhorizontal platforms but not on slopes. According to this model, the accumulation rate of dust was an order of magnitude higher in glacial times than during the interglacials. This conclusion is consistent with a large body of palaeoenvironmental evidence (Liu 1988, 1991; Dodonov 1991; Lazarenko et al. 1981; Pye & Zhou 1989). Thus, despite uncertainties regarding the origin of the susceptibility signal (Zhou et al. 1990; Maher & Thompson 1991) this model provides a better basis for age interpolation than the assumption of uniform dust deposition.

In Figure 2, we present the susceptibility profiles (solid lines) of Karamaidan and Xifeng II on timescales obtained by interpolation of the accumulated susceptibility between the base of PC1 and S1, taken as equivalent to 128 ka, and the B/M boundary at 780 ka. The marine $\delta^{18}O$ record used here is the astronomically tuned SPECMAP record (Imbrie et al. 1984) to 620 ka and planktonic data from ODP Site 677 (Shackleton

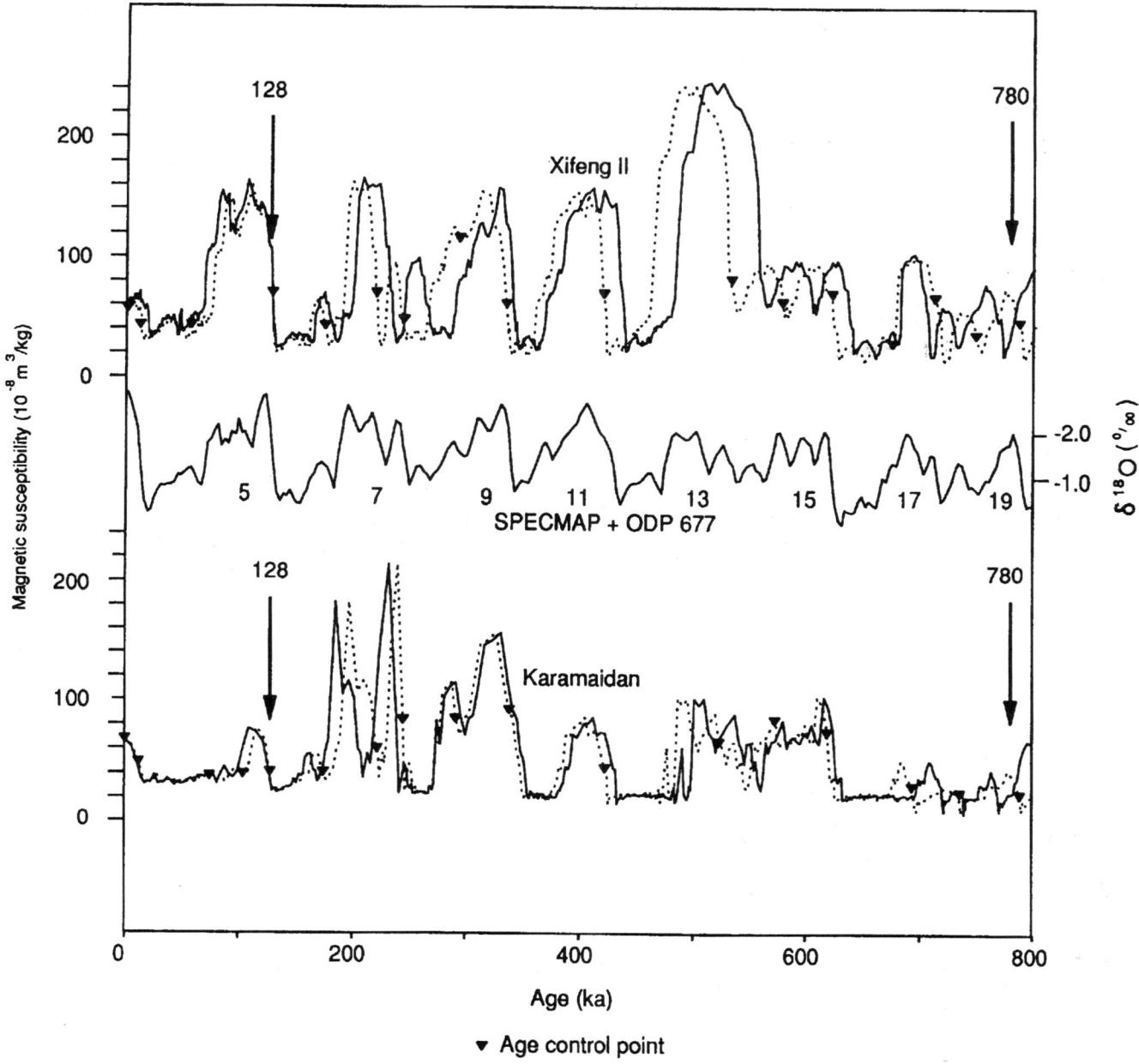

Figure 2. The susceptibility record from Xifeng II (above) and Karamaidan (below) on the untuned (solid line) and tuned (dotted line) susceptibility time scale (ka) compared to the oxygen isotope record of SPECMAP and ODP Site 677 (middle). The two control points, 128 ka and Brunhes/Matuyama boundary (780 ka), are indicated by arrows.

et al. 1990) above that age. The SPECMAP stack, which was published scaled in standard deviation units (Imbrie *et al.* 1984), has been re-scaled to have the same mean and standard deviation as data from the upper part (0 to 620 ka) of the ODP 677 record. The number and relative timing of the major cycles in Karamaidan and Xifeng II and in the oxygen isotope record are similar, although the amplitudes vary. Thus even using this very simple approach, the ages of the Middle Pleistocene soils at both localities correlate reasonably well with those of the interglacials in the marine record. The fact that this approach leads to a reconstructed record of soil development in both China and Tadjikistan that is very similar to the marine $\delta^{18}O$ record suggests that it is broadly correct. An important archaeological implication of this chronology is that the youngest Lower Palaeolithic material found in PC3 is no younger than $\delta^{18}O$ Stage 9 (about 320 ka) and that the Mousterian culture could have appeared during Stage 7 or Stage 5, which is much more in keeping with evidence for the appearance of

the Mousterian in the Near East (Bar-Yosef 1992).

In order to obtain a refined picture of varying dust accumulation rates, we have added detail to the susceptibility-based age model. As is commonplace in the study of deep sea records, we inserted some additional age control points in the susceptibility record (Fig. 2 dotted line) in order to align loess/palaeosol transitions in the magnetic susceptibility record with those in the $\delta^{18}O$ record. This tuning required relatively small changes to the ages obtained by interpolation. For this study we used only a restricted number of control points (16, compared with over 80 in the same time interval in the study by Imbrie *et al.* 1984) which are located at the chief transitions from loess below to soil above, and to analogous minor transitions. It is generally believed that these "Terminations" were rapid and synchronous climatic transitions from glacial to interglacial climates (Broecker & van Donk 1970), so that they provide reliable tie-points for correlation. On the other hand, interglacial-to-glacial transitions are less

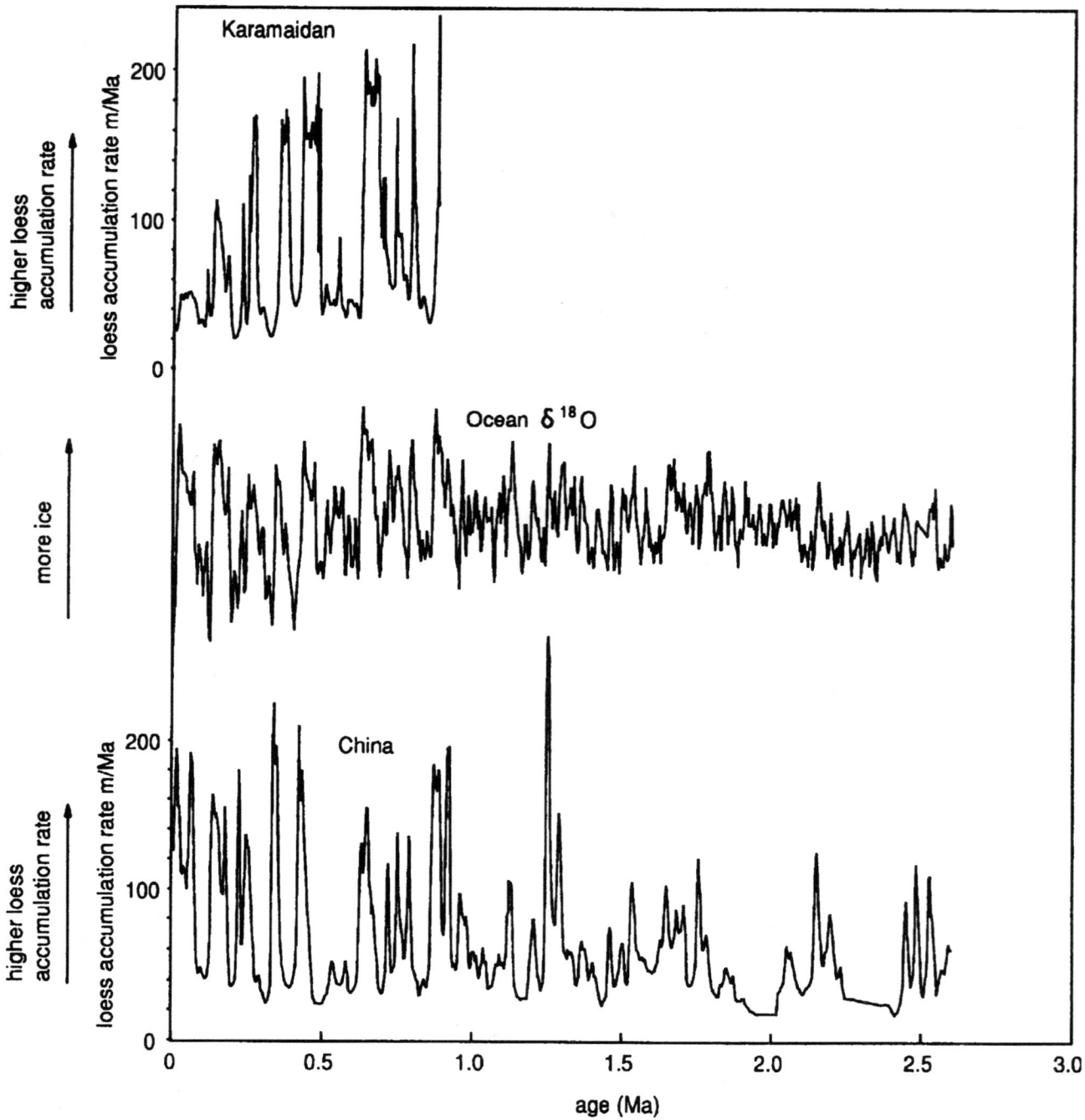

Figure 3. Loess accumulation rates for Karamaidan (above) and China (below), calculated using our tuned age model (Fig. 2, dotted), compared with the δ¹⁸O record (middle) of the SPECMAP Stack (to 620 ka) and ODP 677 (older than 620 ka). Higher dust accumulation rate is associated with increased ice volume.

well-defined.

Loess Accumulation Rates

Using this modified age model, we have calculated the mean sedimentation rate history for the three Chinese records and for the Karamaidan section. As shown in Fig. 3, the similarity between the two records is remarkable, both showing significantly increased dust accumulation rate corresponding to expanded ice volume. This lends support to our initial correlation of the two sections. The only major discrepancy is that the accumulation rate at Karamaidan during the last glacial was rather low according to our model. The reason for this is unclear, but we note that PC1 and the overlying loess are better developed and thicker at several other sections in

Tadjikistan and Uzbekistan: in this respect, Karamaidan is anomalous.

To investigate the relationship between the rate of dust accumulation and the marine δ¹⁸O record, we performed cross-spectral analysis of the dust accumulation rate in Karamaidan (to 0.8 Ma) and the stacked Chinese sections (to 1 Ma) versus δ¹⁸O. The records are highly coherent (0.95) at the 100 ka eccentricity period, and significantly (0.80) coherent at the 41 ka obliquity period (Fig. 4A and 4B). We then compared the dust accumulation rate in the stacked Chinese sections for the entire 2.6 Ma with the δ¹⁸O record (Fig. 4D), and with a theoretical ice volume model (Imbrie & Imbrie 1980) calculated from orbital parameters (Fig. 4C). Given that the time series is tuned to orbital frequencies, it is not surprising that the eccentricity and obliquity periods are present in the spectra.

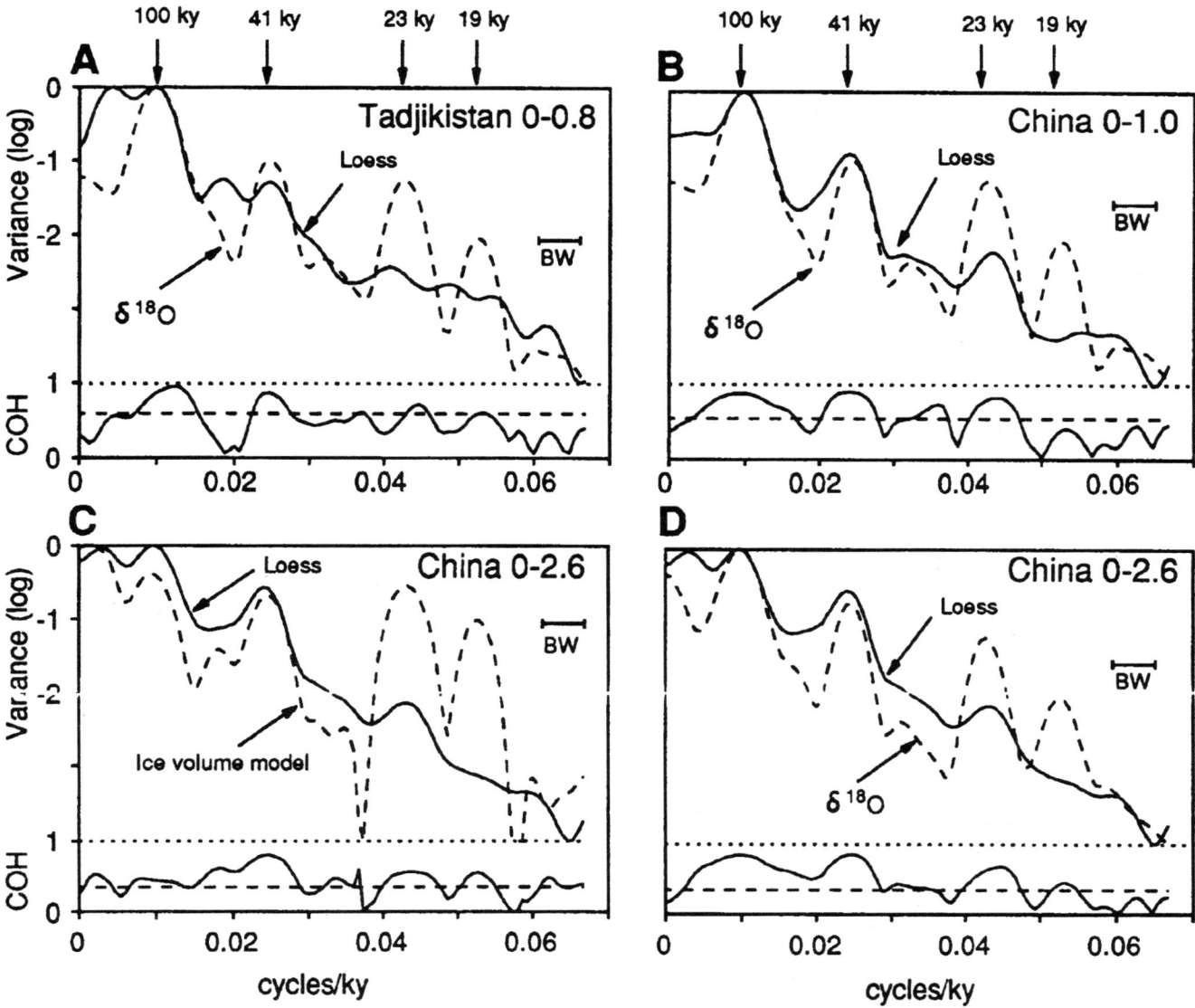

Figure 4. Results of cross spectral analyses (a) of Karamaidan dust accumulation rate (0-0.8 Ma) versus $\delta^{18}O$; (b) Chinese dust accumulation (0-1 Ma) versus $\delta^{18}O$; (c) Chinese dust accumulation (0-2.6 Ma) versus orbital ice volume model; (d) Chinese dust accumulation (0-2.6 Ma) versus $\delta^{18}O$. BW shows bandwidth. COH shows coherency with values above the horizontal line being significant at the 80% level.

However, the very high coherency between the dust accumulation rate and the $\delta^{18}O$, especially at the 100 ka period, is striking. The origin of the eccentricity signal in Quaternary palaeoclimatic records is still debated (Imbrie *et al.* 1993). It is possible that the climate system could respond to eccentricity via direct controls of the strength of the precessional cycle affecting the monsoon, and correspondingly controlling the dust deposition in the middle and low latitudes. However, our results show that at the eccentricity frequency the Pleistocene loess accumulation is much more coherent with the marine $\delta^{18}O$ signal than with the orbitally modelled ice volume or with orbital insolation. This indicates that the major variability in dust accumulation is strongly coupled to continental-scale glaciations rather than directly controlled by insolation forcing.

Acknowledgements

We are very grateful to The Royal Society both for travel grants which have enabled A. E. Dodonov and V. A. Ranov to work in Cambridge, and for support of our research expenses. We thank Simon Crowhurst for assistance with data manipulation.

References

BAR-YOSEF, O. (1992). The role of Western Asia in modern human origins. *Philosophical Transactions of the Royal Society,* Biological Sciences, 337:193-200.

BROECKER, W.S., & VAN DONK, J. (1970). Insolation changes, ice-volumes, and the ^{18}O record in deep-sea cores. *Review of Geophysics and Space Physics.* 8:169198.

CLIMAP project members (1976). The surface of the ice-age earth. *Science,* 191:1131-1137.

CLIMAP project members (1981). Seasonal reconstructions of the earth's surface at the last glacial maximum. *Geological Society of America Map and Chart Series.* Cline, R. (Editor), 36:1-18.

DODONOV, A.E. (1991). Loess of central Asia. *GeoJournal,* 24:185-194.

HELLER, F. & LIU, T.S. (1982). Magnetostratigrphical dating of loess deposits in China. *Nature* 300:431-3.

IMBRIE, J., BERGER, A., BOYLE, E.A., CLEMENS, S.C., DUFFY, A., HOWARD, W.R., KUKLA, G., KUTZBACH, J., MARTINSON, D.G., MCINTYRE, A., MIX, A.C., MOLFINO, B., MORLEY, J.J., PETERSON, L.C., PISIAS, N.G., PRELL, W.L., RAYMO, M.E., SHACKLETON, N.J. & TOGGWEILER, J.R. (1993). On the structure and origin of major glaciation cycles 2. The 100,000-year cycle. *Paleoceanography*, 8:699-735.

IMBRIE, J., HAYS, J.D., MARTINSON, D.G., MCINTYRE, A., MIX, A., MORLEY, J.J., PISIAS, N.G., PRELL, W. & SHACKLETON, N.J. (1984). The orbital theory of Pleistocene climate: support from a revised chronology of the marine $\delta^{18}O$ record. In: Berger, A.L., Imbrie, J., Hays, J., Kukla, G. and Saltzman, B. (Editors), *Milankovitch and Climate*. D. Reidel, Hingham, Mass., Part 1, pp. 269-305.

IMBRIE, J. & IMBRIE, J.Z. (1980). Modelling the climatic response to orbital variations. *Science*, 207:943-953.

KUKLA, G.J. (1977). Pleistocene land-sea correlations. I. Europe. *Earth-Science Reviews*, 13:307-374.

KUKLA, G., HELLER, F., LIU, X., XU, T., LIU, T & AN, Z. (1988). Pleistocene climates in China dated by magnetic susceptibility. *Geology*, 16:811-814.

KUKLA, G., AN, Z.S., MELICE, J.L., GAVIN, J & XIAO, J.L. (1990). Magnetic susceptibility record of Chinese Loess. *Transactions of the Royal Society of Edinburgh*: Earth Sciences, 81:263-288.

LAZARENKO, A.A. BOLIKHOVSKAYA, N.S. & SEMENOV, V.V. (1981). An attempt at a detailed stratigraphic subdivision of the loess association of the Tashkent region. *International Geology Review*. 23:1335-1346.

LIU, T.S. (editor) (1988). *Loess and the Environment*. Springer, Berlin 251 pp.

LIU, T.S. (editor) (1991). *Loess, Environment and Global Change*. Science Press, Beijing. 288 pp.

MAHER, B. & THOMPSON, R. (1991). Mineral magnetic record of the Chinese loess and paleosols. *Geology*, 19:3-6.

PEN'KOV, A.V. (1982). In: Dodonov, A.E. (Editor), *Guidebook for international field excursions* A-11, C-11. XI INQUA Congress, Moscow, pp. 31-66.

PYE, K. & ZHOU, L.P. (1989). Late Pleistocene and Holocene aeolian dust deposition in North China and the Northwest Pacific Ocean. *Palaeogeography, Palaeoclimatology, Palaeoecology* 73 11-23.

RANOV, V. (1988). *The Earliest Pages of Humanity*. Education Press, Moscow (in Russian).

RANOV, V. & DAVIS, R. (1979). Towards a new outline of the Soviet Central Asia Palaeolithic. *Current Anthropology*, 20:249-270.

SHACKLETON, N.J., BERGER, A. & PELTIER, W.R. (1990). An alternative astronomical calibration of the Lower Pleistocene timescale based on ODP Site 677. In: The Late Cenozoic Ice Age. *Transactions of The Royal Society of Edinburgh: Earth Sciences*, 81:251-261.

WINTLE, A.G. & HUNTLEY, D.J. (1982). Thermoluminescence dating of sediments. *Quaternary Science Reviews* 1:31-53.

ZHOU, L.P., OLDFIELD, F., WINTLE, A.G., ROBINSON, S.G. & WANG, J.T. (1990). Partly pedogenic origin of mineral magnetic variations in Chinese loess. *Nature*, 346:737-739.

Quaternary Proceedings No. 4, 1995 7-18
© Quaternary Research Association, Cambridge.

Loess-Palaeosol Sequences as Recorders of Palaeoclimatic Variations During the Last Glacial-Interglacial Cycle: Some Problems of Correlation in North-Central China.

Edward Derbyshire, David H. Keen, Rob A. Kemp, Tim A. Rolph, John Shaw and Xingmin Meng.

Derbyshire, E., Keen, D.H., Kemp, R.A., Rolph, T.A., Shaw, J. and Meng, X., 1995 Loess-palaeosol sequences as recorders of palaeoclimatic variations during the last Glacial-Interglacial cycle: some problems of correlation in north-central China, In *Wind Blown Sediments in the Quaternary Record* (Edward Derbyshire). Quaternary Proceedings No. 4, John Wiley & Sons Ltd., Chichester, pp. 7- 18.

Abstract

Detailed sampling of several sections in the loess-palaeosol succession of North China has been undertaken along a present-day precipitation and temperature gradient from the humid south-east of the Loess Plateau to its semi-arid western margins. Data on magnetic susceptibility, granulometry, sediment fabric, mineralogy, micromorphology, snails, carbonates and organic carbon were obtained from several sites. The data from these sections show them to be a high resolution record of climatic variation. However, a number of problems affecting the representativeness of the data have been encountered. The advantages provided by high accumulation rates (high resolution record) of the western margins of the Loess Plateau may be offset by 'cut and fill' disparities between adjacent sections. Some lithostratigraphic sequences from which proxy measures of climate have been derived are incomplete, as shown by erosion surfaces, water-laminated zones and truncated palaeosol profiles. In the more humid eastern area of southern Shaanxi Province, a further problem is posed by the clear evidence of superimposition (or "welding") of successive palaeosols, substantial pedogenic modification of the intercalated loessic units, and massive lowering of parts of the plateau surface by human action in historical times. The presence of such features, together with the common stratigraphical practices of 'counting down from the top' and correlating individual loess-palaeosol sections with the marine and the ice core oxygen isotope curves, has implications for stratigraphical correlation across and well beyond the Chinese Loess Plateau.

KEYWORDS: magnetic susceptibility; micromorphology; sedimentary properties; Loess Plateau.

Derbyshire, E., Kemp, R.A., and Meng, X., Centre for Quaternary Research, Department of Geography, Royal Holloway, University of London, Egham, Surrey TW20 0EX, UK.

Keen, D.H., School of Natural and Environmental Science, Coventry University, Priory Street, Coventry CV1 5FB, UK.

Rolph, T.A. and Shaw, J., Quaternary Environmental Research Centre, University of Liverpool, Oliver Lodge Building, Liverpool L69 3BX, UK.

Introduction

The Loess Plateau of northern China lies between the rivers Hwang (Yellow) and Wei, and covers an estimated area of 275,600 km² (Liu *et al.* 1964: see Fig. 1). It consists of wind-blown silts often over 100m thick. Exceptional thicknesses of more than 300m are known near the city of Lanzhou (Derbyshire 1984). The potential value of these loess sequences, with their many intercalated palaeosols, as a record of changing Quaternary climates was first recognised in the late 1950s (Liu 1958). On the basis of fossil assemblages, Liu (1964) demonstrated that deposition of the Chinese loess extended throughout the entire Quaternary. The early work of Liu and Chang (1964) concentrated on the type site at Luochuan in

Shaanxi Province to establish the first-order loess-palaeosol lithostratigraphy of the Loess Plateau. This work refined the original succession consisting, from oldest to youngest, of Wucheng, Lishi and Malan loess. Alternating palaeosol and loess units were assigned 'S' and 'L' labels respectively, and each numbered consecutively from the top downwards (S_0, L_1, S_1, L_2, etc.).

Studies of palaeomagnetic variations in the loess at Luochuan established an age for the basal loess contact with the Pliocene red "clay" (the Luochuan Red Loam Formation: Liu & Yuan 1987) of 2.48 Ma (Heller & Liu 1982,1984). The Olduvai Sub-Chron was fixed at a depth of 100-108m (coincident with WS_1, the sixteenth buried palaeosol below S_0), and the Jaramillo Sub-Chron at 66-70m (between loess

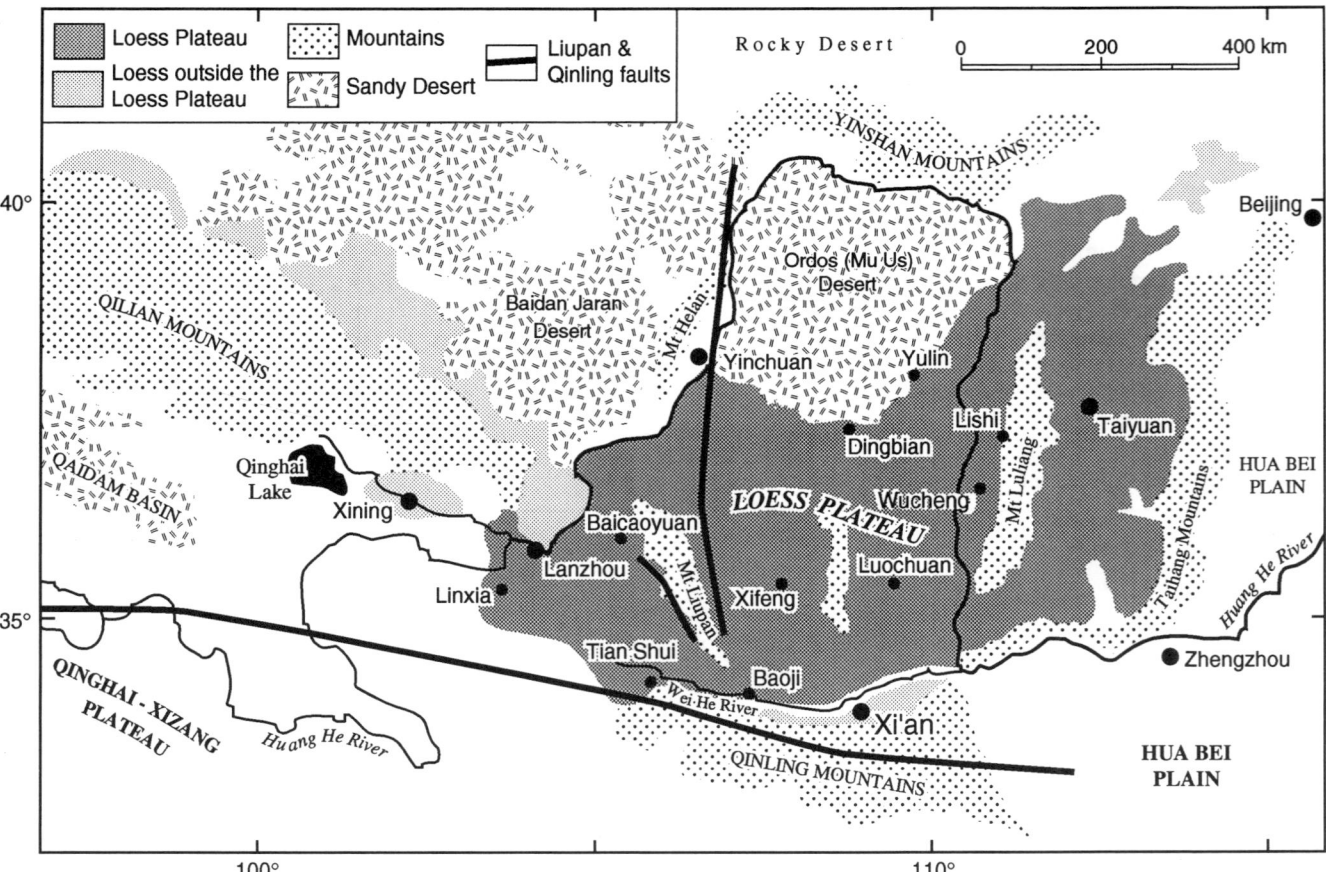

Figure 1. The Loess Plateau of China and its environs.

units L_{10} and L_{11}, and including S_{10}). The Brunhes-Matuyama transition was recognised at a depth of 52m, near the top of palaeosol S_8. Distinctly sandy loess units, notably L_9 and L_{15} (Liu 1985; Liu et al. 1985), were interpreted as representing periods of advancing desert margins that imply movements of regional climatic boundaries by up to 500 km. Using a combination of loess weathering status, Fe_2O_3/FeO ratios, grain size, carbonate content, and palaeosol typology, Liu et al. (1985) constructed curves for mean annual temperature and precipitation, and shifts in forest-steppe to desert-steppe at Luochuan for the past 2.4 Ma.

The Loess Plateau is a region of steep climatic gradients, with conditions ranging from semi-arid in the north and west, to sub-humid in the south and east. The climate is monsoonal in that there are severe seasonal shifts in dominant winds. In winter and spring dry, cold winds from the north and west are induced by the Mongolian-Siberian high and the Aleutian low pressure systems (Zhang & Lin 1992). In contrast, summer is marked by the ingress of warm, oceanic air derived mainly from the South China Sea with some influx from the Bay of Bengal. Using the present climatic regime as a model, the loess-palaeosol sequence in the Loess Plateau has been widely interpreted (e.g. Liu & Wang 1964; Liu 1985; Liu et al. 1985; Li & Feng 1988; An et al. 1991a, 1991b; Zhang 1992; Ding et al. 1992; Liu & Ding 1993) in terms of cold, windy conditions during glacials under the influence of a dominantly north-westerly airflow (i.e. winter pattern), and milder, moister conditions in interglacials with a relatively greater influence of south-easterly winds from the South China Sea (summer pattern). Readily-derived data, most notably field magnetic susceptibility and particle size, have been used to provide a first-order indication of the relative duration and strength of the two

principal components of the Asian monsoon system (e.g. An et al. 1991a, 1991b; Ding et al. 1992).

Low-field magnetic susceptibility profiling has been widely used to provide a high resolution framework at many sites in the loess terrain of North China. Systematic variations in magnetic susceptibility are a function of variable ferrimagnetic (magnetite and maghaemite) mineral grain-size and concentration, these being coarser and lower, respectively, in the loess compared to the palaeosols (Heller & Liu 1986; Kukla 1987; Kukla et al. 1988; Kukla & An 1989; Heller et al. 1991). Analyses of the magnetic mineralogy in different parts of the Loess Plateau (Zheng et al. 1991; Zhou et al. 1990) have indicated that the fluctuations in the magnetic susceptibility curves are strongly controlled by a the in situ modification of magnetic minerals rather than by a simple variation in the dilution (by incoming loessic dust) of a constant 'magnetic rain' (Kukla 1987). The data from the Lanzhou loess show that the increased concentration of ferrimagnetic minerals in palaeosol units consists not only of extra-fine grains (<100nm) but also coarse grains (>100nm). The superparamagnetic (SP) grains (<0.03μm) are likely to have had a different origin from the coarser single domain (SD) and multidomain (MD) grains. This indicates that, although the in situ formation of SP ferrimagnets dominates the enhanced low-field magnetic susceptibility of palaeosols, an enhancement of the coarser fraction through leaching also plays an important role during pedogenesis (Liu et al. 1994).

There is no doubting the inter-regional similarity between many magnetic susceptibility curves and curves of other climate-proxy measures derived from the Chinese loess sections such as particle size parameters (including, inter alia, median grain size, percentage clay content and clay-silt ratios). Median grain size, and various grain size ratios, are now generally regarded as a measure of wind strength in conditions of rapid sedimentation, as exemplified by the thick loess-palaeosol series of Central and East Asia (*e.g.* Liu *et al.* 1985; Ding *et al.* 1992; Liu & Ding 1993). Specifically, median grain size is regarded as an expression of the vigour of the north-westerly (winter) monsoon, as exemplified by studies of the dynamics of the present-day system (Liu *et al.* 1989). It is argued that median values are distinctly coarser in the cold and dry glacials compared to the warmer and moister interglacials, with their enhanced south-easterly monsoonal winds and relatively suppressed north-westerlies. In the case of the very high resolution sections around Lanzhou, the degree of detail provided by the particle size parameters is greater than that derived from the magnetic susceptibility curves. This is because acquisition of a magnetic susceptibility signal is a long-term process (taking at least several decades) that is unlikely to reflect short-term events. Grain size, on the other hand, is a primary property of loess that responds instantaneously to variations in wind strength.

Magnetic susceptibility profiling has been adopted as a standard technique in intra- and inter-regional stratigraphical correlation and, increasingly, as a measure of the variability of the palaeomonsoonal system of eastern Asia (An *et al.* 1991a, 1991b, 1993: Porter *et al.* 1992). Magnetic susceptibility curves from an increasing number of individual sites are being correlated with distant sites elsewhere in the Loess Plateau. Results from individual loess-palaeosol sites are also being correlated with the astronomically-tuned Oxygen isotope (SPECMAP) record (Imbrie et al. 1984) from the deep oceans (*e.g.* Liu & Yuan 1982; Kukla *et al.* 1990) and the isotopic curves from the Vostok ice core (Li *et al.* 1992). Such comparative studies led Ding *et al.* (1992) to propose a coupling mechanism between global ice volumes and the behaviour of the palaeomonsoons, the record of the latter clearly bearing the marks of a stepwise uplift of the Tibetan Plateau (Liu & Ding 1993). It is important to recognise, however, that correlation of the magnetic record in the loess with the deep ocean Oxygen isotope record has sometimes proved to be a tortuous operation as, for example, in the case of the classic loess-palaeosol site at Luochuan (Heller & Liu 1986). In much of the work so far completed in North China, established practice has been to read the lithostratigraphy by 'counting from the top', and correlating with the stratigraphy provided by the regional stratotype at Luochuan (Liu 1985). Thus far, relatively little consideration has been given to the implications for this methodology of regional and local variations in the loess-palaeosol record. Correlation between individual loess sites may appear disarmingly simple, especially using conventional field counting, magnetic susceptibility and particle size curves in the absence of datable material.

A number of problems concerned with the establishment and correlation of stratigraphy were encountered at several sites during the course of a multi-disciplinary study of a selection of proxy-measures of climate along a WNW-ESE climatic gradient in the southern Loess Plateau from near Xian in the east to Lanzhou in the west. These are discussed here, partly in order to clarify the stratigraphy at each site but mainly to illustrate the potential difficulties encountered in correlation across the Loess Plateau, as well as to underline the importance of validating the local stratigraphy prior to such correlation. The sites described are Jiuzhoutai and Gaolanshan, to the north and south respectively of Lanzhou city on the semi-arid western margins of the loess region, and Liu Jia Po, east of the city of Xian in the humid, warmer southern part of the Loess

Plateau. Mean annual precipitation is 328mm at Lanzhou and 650mm at Liu Jia Po. Closely spaced samples were obtained for study of the grain size, magnetic susceptibility, sediment fabric, micromorphology, mineralogy, Mollusca, carbonates, and organic carbon. The ages of the material were estimated using palaeomagnetic methods at Lanzhou and luminescence techniques at Liu Jia Po.

Jiuzhoutai, Lanzhou

Jiuzhoutai Mountain (2067m above sea level) lies a few km north of the city of Lanzhou between the Tibetan Plateau and the western margins of the Loess Plateau (Fig. 1). Major fault systems, notably the WNW-ESE Qinling Shan and NNW-SSE Liupan Shan groups, are formative features in the current uplift of the Tibetan Plateau. Earthquake shock is frequent (Dijkstra *et al.* 1993). Moreover, a highly seasonal rainfall regime, in which up to 75% of the annual rainfall total may fall in a few days in summer under the influence of the south-easterly monsoon, generates overland flow which is diverted down the abundant fissures in the loess to create a sink-and-pipe system known as "loess karst" (Muxart *et al.* 1994). The result is a landscape in which the upper few metres of the loess slopes suffer repeated collapse, reworking and landsliding (Derbyshire *et al.* 1991).

The Hwang He (Yellow River) and its tributaries have responded to the Tibetan Plateau uplift by a deep incision that has produced a mountainous terrain with a relative relief of up to 600m, and steep, rectilinear slopes with gradients of between 28-40°. The upper slopes of Jiuzhoutai Mountain (2067m above sea level), consist of the thickest recorded loess-palaeosol

Figure 2. One of the deeper parts of the trench section on Jiuzhoutai Mountain in 1984. Note the precipitous slopes in the loess (here 39°). View south from about 50m above the loess/bedrock contact.

Figure 3. Laminated unit *c.* 0.5m thick in the Jiuzhoutai trench section, about 80m above the base.

sequence in the world (315m). The first magnetostratigraphical work was completed here in the early 1980s (Wang 1982) and several studies by other groups have since been undertaken. All studies published to date, however, have been based on samples derived from a shallow trench cut into a steep (35-39°) slope on the south face of the mountain (Fig. 2), which contains features such as shear planes (down to depths of a few metres), erosion surfaces and thin zones of laminated silts (Fig. 3).

The basal age of the loess column at Jiuzhoutai has been variously estimated, using palaeomagnetic data, as ~1.3Ma (Burbank & Li 1985) and ~2.4 Ma (Rolph *et al.* 1989), although both these studies agreed that the Brunhes-Matuyama polarity boundary lies between 144 and 148m above the base. Estimates of the depth below the present land surface of palaeosol S_1

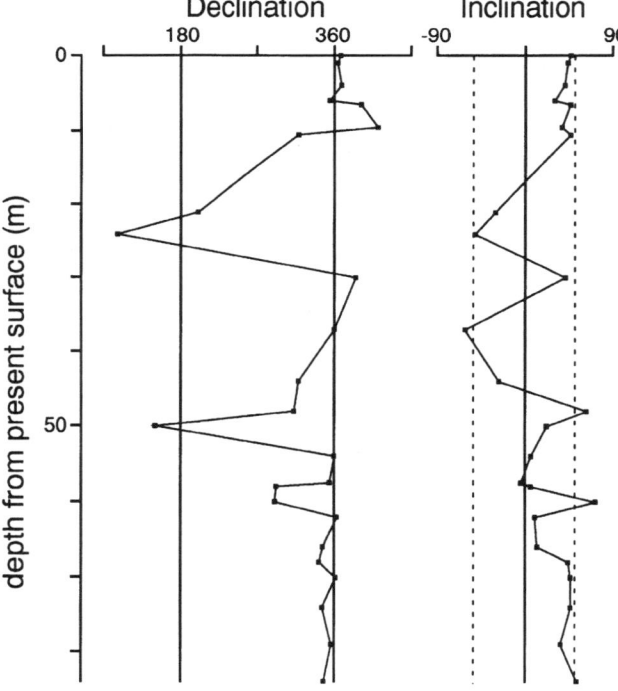

Figure 4. Palaeomagnetic record of the uppermost 85m of the Jiuzhoutai loess section. The directions within the depth interval 37-70m are incompatible with the known geomagnetic behaviour in the Late Quaternary. The vertical broken lines on the Inclination plot represent the expected magnetic inclination for a stable geomagnetic field of normal or reversed polarity.

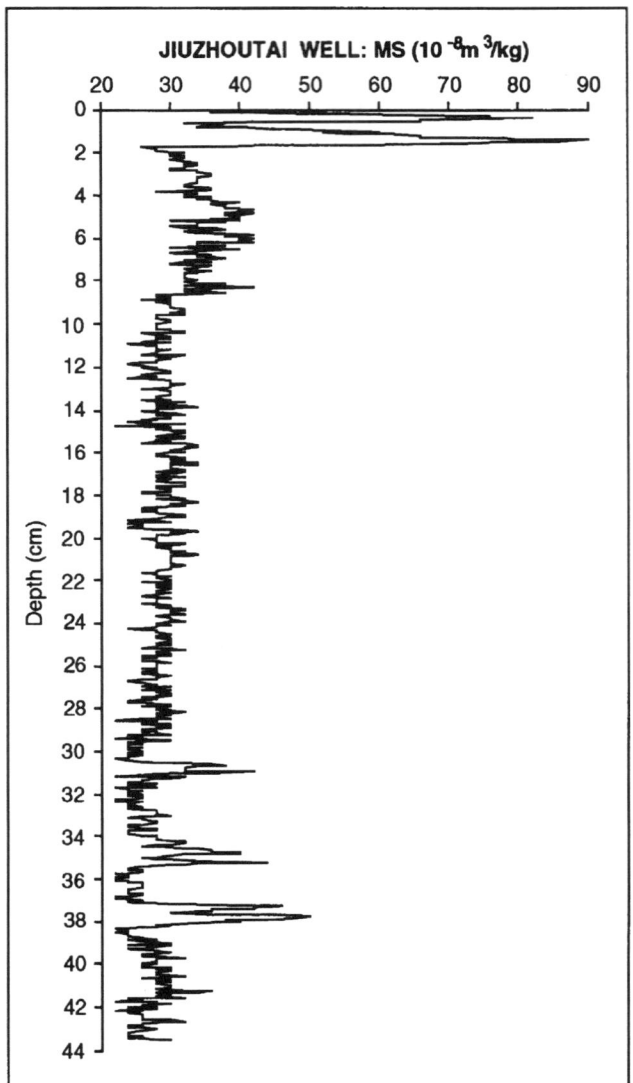

Figure 5. Magnetic susceptibility curve for the Jiuzhoutai well section, Lanzhou.

(widely correlated with Oxygen isotope stage 5, *e.g.* Liu 1985; Kukla *et al.* 1990; Liu & Ding 1993) have varied from ~65m (Burbank & Li 1985) to 58-64m (Chen *et al.* 1991). The palaeomagnetic study of the trench by Rolph *et al.* (1989), while noting the many discrete erosion surfaces and shear planes, produced a detailed magnetostratigraphic log from which it was possible to define the Brunhes-Matuyama boundary and a number of normal polarity events within the Matuyama reversed polarity Chron. However, interpretation of the upper part of the section proved difficult in detail. It was found to contain a series of palaeomagnetic directions at depths of between 37 and 70m (Fig. 4). Using the calculated average accumulation rate of 25cm/ka based on the Brunhes-Matuyama boundary (age 780 ka) at a depth of 195m (Rolph *et al.* 1989), these fluctuations suggested that the magnetic field had been in an unstable state for a time interval of approximately 132 ka, from 280 ka to 148 ka. While it is known that a number of geomagnetic excursions occurred within the Late Quaternary (*cf.* Blake Event and Lake Biwa events), the duration of such events (and even of full polarity reversals) is of the order of 10 to 20 ka. Such an apparently continuous record of unstable polarity cannot be reconciled with the Late Quaternary geomagnetic field record: rather, it is suggested that the upper part of the section has suffered physical disturbance in the recent past.

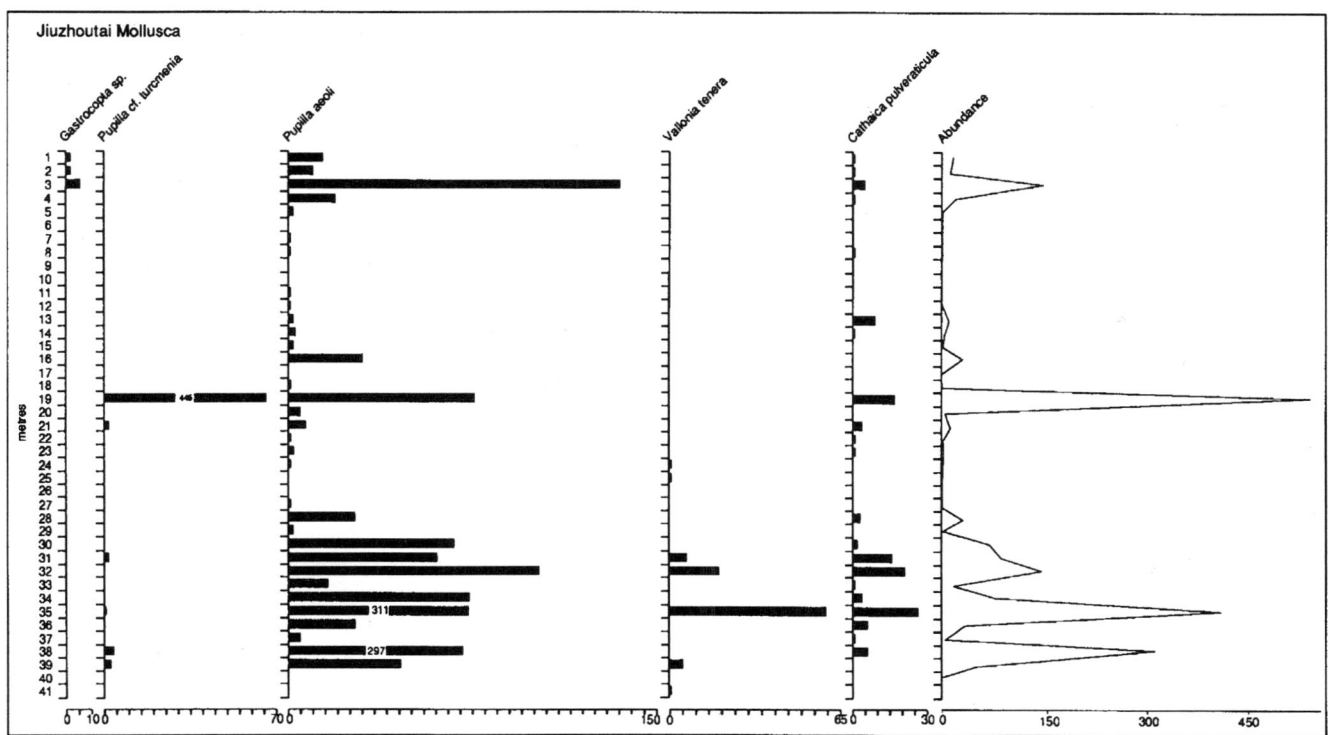

Figure 6. The molluscan fauna from the Jiuzhoutai well section. Counts are of absolute numbers. The combined curve to the right of the diagram lists total molluscan abundance.

By 1993, it was clear that further advance in the study of the lithostratigraphy of the very thick loess-palaeosol sequences around Lanzhou demanded access to exposures well away from the actively-degrading slopes. To this end, Professor Wang Jingtai agreed to assist the authors by having a vertical well, 2m in diameter and 44m deep, excavated on the flat summit of Jiuzhoutai Mountain. This hand-dug well, situated more than 10m from the nearest slope, has provided the authors with a closely-spaced set of field measurements of magnetic susceptibility, as well as a semi-continuous set of oriented and levelled block samples each 10 x 10 x 10cm in size. Bulk sediment samples at 1m vertical intervals, and weighing up to 10kg, were taken for molluscan analysis.

Magnetic susceptibility was measured at 5cm intervals on the block samples from the Jiuzhoutai well. The seventy measurements taken between 28 and 40m below the surface show three susceptibility maxima between c. 30m and 38m (Fig. 5). These have a general form similar to triple peaks interpreted at several sites across the Loess Plateau as indicating S_1, including Dawan (30 km west of Lanzhou: Rolph et al. 1993), and Beiyuan (near Linxia) where luminescence ages of 63.8±5.9 and 141.4±14.1 ka bracket the palaeosol complex (An et al. 1991a).

Luminescence dating of the block samples from the well section on Jiuzhoutai is not yet complete. However, the contained molluscan faunas have been studied. The bulk samples were sieved to 500µm. Mollusc counts followed the conventions of Sparks (1963), each complete shell or unique part of a shell counting as one individual. The results are shown as a histogram (Fig. 6). It is evident that the Jiuzhoutai section is relatively poor in molluscan remains, with a total of only five taxa having been found. Much the most important species is *Pupilla aeoli* (Hilber), with *Cathaica pulveraticula* (von Martens) being the next most notable, followed by *Vallonia tenera* (Reinhardt) and a Pupillid tentatively identified as *Pupilla turcmenia* (Boettger). The molluscan numbers peak significantly

at only five levels: 3m, 19m, and at depths corresponding to the three magnetic susceptibility peaks in the complex between c.28m and 38m. Because the major controls on mollusc existence are humidity and temperature, it is probable that these peaks represent episodes of warmer and more humid climates in this region, and that the levels essentially devoid of such fauna represent cold and dry periods. The large molluscan numbers at a depth of 19m consist overwhelmingly of *Pupilla turcmenia*, a species which lives in Gansu at present, but is also known at altitudes of 4000m in Tibet and at sites in Central Asia (Likharev & Rammel'meier 1952). This hardy species is probably well adapted to take advantage of a climatic fluctuation which was both short-lived and of shallow intensity. Such an interpretation is consistent with magnetic susceptibilty readings suggesting only very weak pedogenesis at this depth (S_m) on Jiuzhoutai (Fig. 5) and at stratigraphically-equivalent depths on Gaolanshan (see below), although substantiation must await independent dating. This climatic fluctuation is here tentatively correlated with Oxygen Isotope Stage 3. The triple peaks between 28m and 38m are typified by significant numbers of *P. aeoli*, *C. pulveraticula* and the appearance of *Vallonia* and *Gastrocopta*. The presence at these three magnetic susceptibility maxima of *Vallonia tenera*, a species that is notably abundant in the loess succession in the much warmer and more humid southern region Shaanxi Province to the east (as shown by our results from Liu Jia Po, discussed elsewhere), indicates considerably milder conditions punctuated by two cold and dry phases. These data, taken together with the palaeopedological evidence of warm and moist climatic phases at similar depths on Gaolanshan (see below), could be taken as indicating that the triple peak in the magnetic susceptibility profile from our well section on Jiuzhoutai represents S_1 and sub-Stages 5e, 5c and 5a in the Oxygen isotope chronology.

This recent work on Jiuzhoutai has raised a number of systematic problems, in particular the difficulties introduced into stratigraphic correlation by discontinuities caused by

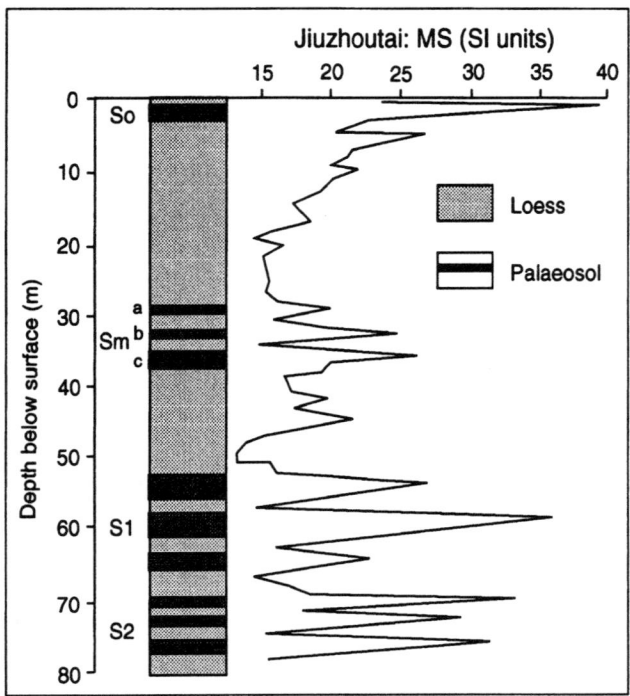

Figure 7. Part of the lithostratigraphical and magnetic susceptibility log for the trench section on Jiuzhoutai, re-drawn after Chen et al. (1991). It shows the last interstadial soil (S_m) at between 32 and 38m depth, and the last interglacial soil (S_1) at 58 - 64m.

mass movements, features which are often difficult to detect in massive, homogeneous sediments such as loess. Other problems arise from the practice of counting loess and palaeosol units from the top and 'matching' the magnetic susceptibility curves with those of the type stratigraphy at Luochuan, hundreds of kilometres away. The interpretation of the Jiuzhoutai magnetic susceptibility by Chen et al. (1991) provides a case in point (Fig. 7). It appears that palaeosol S_1 was identified at a depth

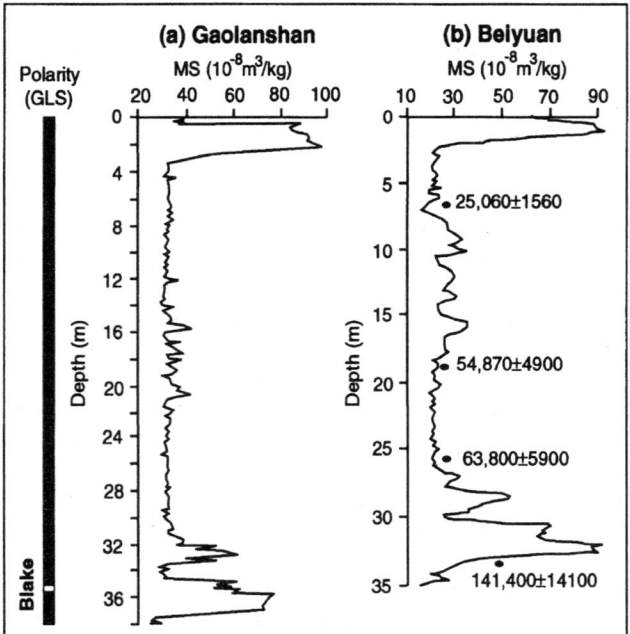

Figure 8. Magnetic susceptibility curves for (a) well 'C' on Gaolanshan, Lanzhou, and (b) Beiyuan, near Linxia, south of Lanzhou, after An et al. (1991b).

(58 - 64m) at which it was *expected* on the grounds that S_1 is a considerably more developed soil than S_2 at the classic Luochuan site. However, the converse may be true in the western (Gansu Province) loess region. Comparison of our unpublished magnetic susceptibility curves for Lanzhou with that for Xifeng (Fig. 7 in Rutter 1992), and the finer median grain size and higher clay contents in S_2 compared to S_1 around Lanzhou, suggest that S_2 may be more developed than S_1 in the western parts of the Loess Plateau. The complex of three peaks at 32-38m was assigned interstadial status by Chen et al. (op. cit.) and mistakenly termed S_m rather than S_1.

Gaolanshan, Lanzhou

Gaolanshan ('shan' = mountain) has a summit altitude of 2135m, and lies some 10km SSE of Jiuzhoutai on the opposite (southern) side of the Yellow River. In 1993, a hand-dug well was excavated on the flat summit of the mountain. Excavation of vertical wells to depths of tens of metres is an expensive and hazardous operation, even in relatively soft and dry materials such as loess. For this reason, the depth excavated will always be minimised, the decision to terminate depending upon on-site interpretation. This may be a matter of fine judgement depending, as it must, on interpretation of rapid magnetic susceptibility readings and evaluation of loose samples hoisted to the surface.

The well, known as well 'C', extended down to a depth of 38m. It provided a semi-continuous set of oriented and levelled block samples 10 x 10 x 10cm in size, at intervals of 20cm. A low-frequency magnetic susceptibility profile was completed at the same time, using a Bartington MS2 meter at vertical intervals of 5cm on the wall of the well. This shows (Fig. 8a), that a triple magnetic susceptibility peak occurs at depths of between about 32 and 38m. On comparison with the magnetic susceptibility record at Beiyuan (Fig. 8b), this is considered to be palaeosol complex S_1, equivalent of Oxygen Isotope sub-Stages 5e, 5c and 5a. Kemp et al. (1995) interpreted the weakly developed series of palaeosols between 15 and 21m as the S_m complex (Li et al. 1992) correlated with Oxygen Isotope Stage 3 by An et al. (1991a). The mid-Holocene palaeosol (S_0) was missing at well 'C', presumably because of erosion, but is found some 1 to 3m below the surface at well 'A' and in adjacent road-side exposures (Fig. 9)

The decision to terminate the excavation of well 'C' at a depth of 38m was taken on the grounds that the base of S_1 had been penetrated, using as a guide the magnetic susceptibility curve from Beiyuan, Gansu Province (Fig. 8b) and the evident close similarity between these two curves (Fig. 8). The semi-continuous block samples provided the basis for a very detailed pedosedimentary and palaeoenvironmental reconstruction of the S_1 palaeosol based on micromorphology (Kemp et al. 1995). Three main 'soil-forming intervals' were identified, each representing a phase of enhanced pedogenesis on a relatively stable land surface under a climate sufficiently moist to encourage weak leaching. A magnetic reversal was found at 36m (Fig. 8a) and is thought to be the Blake geomagnetic event (115 ka). Kemp et al. (1995) thus tentatively correlated the three soil-forming intervals with Oxygen Isotope sub-Stages 5e, 5c and 5a.

However, other well sections have since been dug on Gaolanshan (Wang Jingtai, personal communication). Comparative study of the magnetic susceptibility data from an adjacent well (well 'B': Fig. 9) now suggests that correlation of individual palaeosols and loessic units is not a simple exercise, even over horizontal distances of as little as 70m. A number of

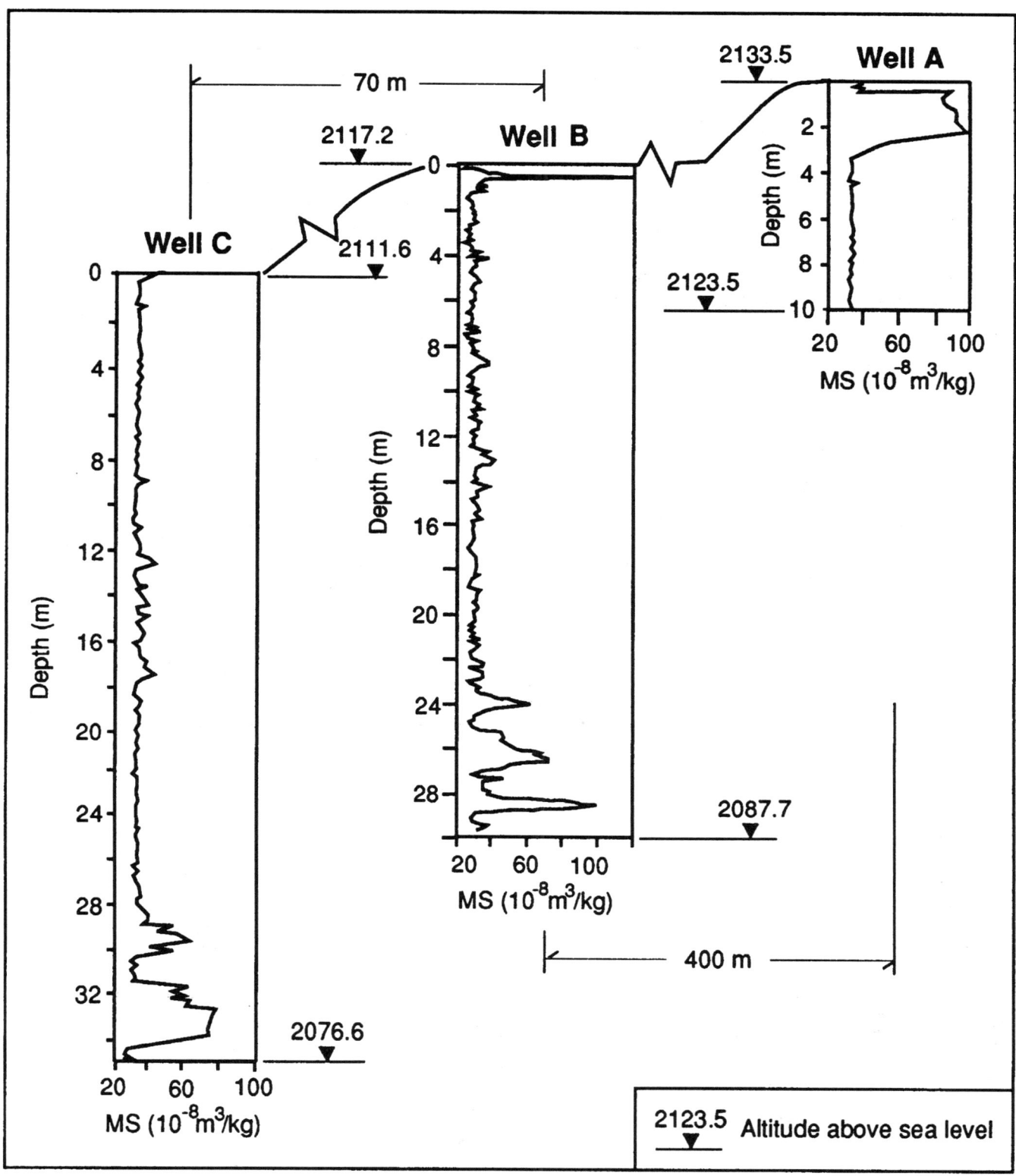

Figure 9. Magnetic susceptibility curves for three adjacent well sections on Gaolanshan, Lanzhou.

differences between these two magnetic susceptibility curves are evident. Apart from the obviously higher overall resolution in well 'C', the zone of weak pedogenesis in mid-section is either compressed or truncated in well 'B' (12.6-14.5m) compared to well 'C' (15.5-21m). The most striking difference, however, lies in the fact that a sharp, discrete susceptibility peak occurs around 28.5m in well 'B', about 2m below the single and double peaks similar to those between 31.5 and 37.5 in well 'C'. Of course, it is possible that such differences reflect short-range variations in the amount of silt deposited or

local erosion and reworking, *i.e.* local changes in the deposition/erosion balance (both cut and fill) during accretion. Equally, however, they may indicate that the decision to cease excavating well 'C' at a depth of 38m was taken marginally too soon, so that the lowermost susceptibility peak was missed. Unfortunately, no block samples were provided to the authors from well 'B'. A series of ten samples from the 'C' well is currently being dated by luminescence at the Physical Research Laboratory in Ahmedabad, India, and these results will be reported elsewhere.

Liu Jia Po

The loess-palaeosol sequence near the village of Liu Jia Po, near Xian, is approximately 110m thick. It rests on alluvial terrace gravels of the Ba River, a tributary of the Wei He which is the principal affluent of the Yellow River.

In the summer of 1993, road-side sections and part of the cliff beneath them were excavated to a total depth of 18m. The lithostratigraphy consists of a thin, greyish surface soil overlying a yellow-brown loess (Munsell 10 YR) with several clearly visible brown to reddish-brown palaeosols (Munsell 7.5 YR). A magnetic susceptibility profile was established in the field and was subsequently confirmed and refined by measurements taken in the laboratory using sets of block samples. The resultant curve (Fig. 10a) shows five maxima at depths of 0.8-2.5m, 4.5-7m, 9-11m, 12.5-14m, and 15.5-18m. This record is very different from those found further west near Lanzhou (Figs. 6 and 8). The peaks shown in Figure 10a have been interpreted by the local geologists, once again, by analogy with the Luochuan type section and counting downwards from the top, as representing the S_0 (mid-Holocene: depth 0.8-2.5m), S_1 (depth 4.5-7m) and S_2 (depth 15.5-18m) palaeosols with two interstadial 'steppe' soils interbedded between S_1 and S_2 (depth 9-11m and 11.5-14m).

The equivalent magnetic susceptibility curve from Duan Jia Po, a site less than 7 km from Liu Jia Po, however, shows six main peaks compared to the five measured at Liu Jia Po (Fig. 10b). Comparison of the curves reveals a strong similarity in both the form and the maximum measured values of the three lowest peaks at the two sites. The fourth peak above the base at Duan Jia Po is of similar magnitude to the fourth peak from the base at Liu Jia Po, but the resolution of this dual peak is clearer at Duan Jia Po. It is in the upper parts of the two curves, however, that a significant difference is evident. The penultimate peak at Duan Jia Po has a very similar magnitude to the uppermost peak at Liu Jia Po (210-225 x $10^{-8}m^3$/kg). A peak equivalent to the uppermost maximum at Duan Jia Po could not be found at Liu Jia Po, despite running a series of magnetic susceptibility profiles at three different points along a 40m section of this exposure. This again shows the difficulty in using magnetic susceptibility curves for correlation, even over distances of a few km.

In the drier Lanzhou region to the west, maximum carbonate values (14-18%) occur up to 40cm below the magnetic susceptibility peaks, a reflection of the level to which solutes are leached and reprecipitated during major phases of soil formation (Kemp *et al.* 1995). Detailed studies of the upper 8m at Liu Jia Po have been carried out. Both the magnetic susceptibility and the median grain size records discriminate clearly between the 'loess' and 'palaeosol' units (Fig. 11a, b). However, the buried soils all overlap each other, as shown by the distribution of the illuvial clay in thin sections and also by

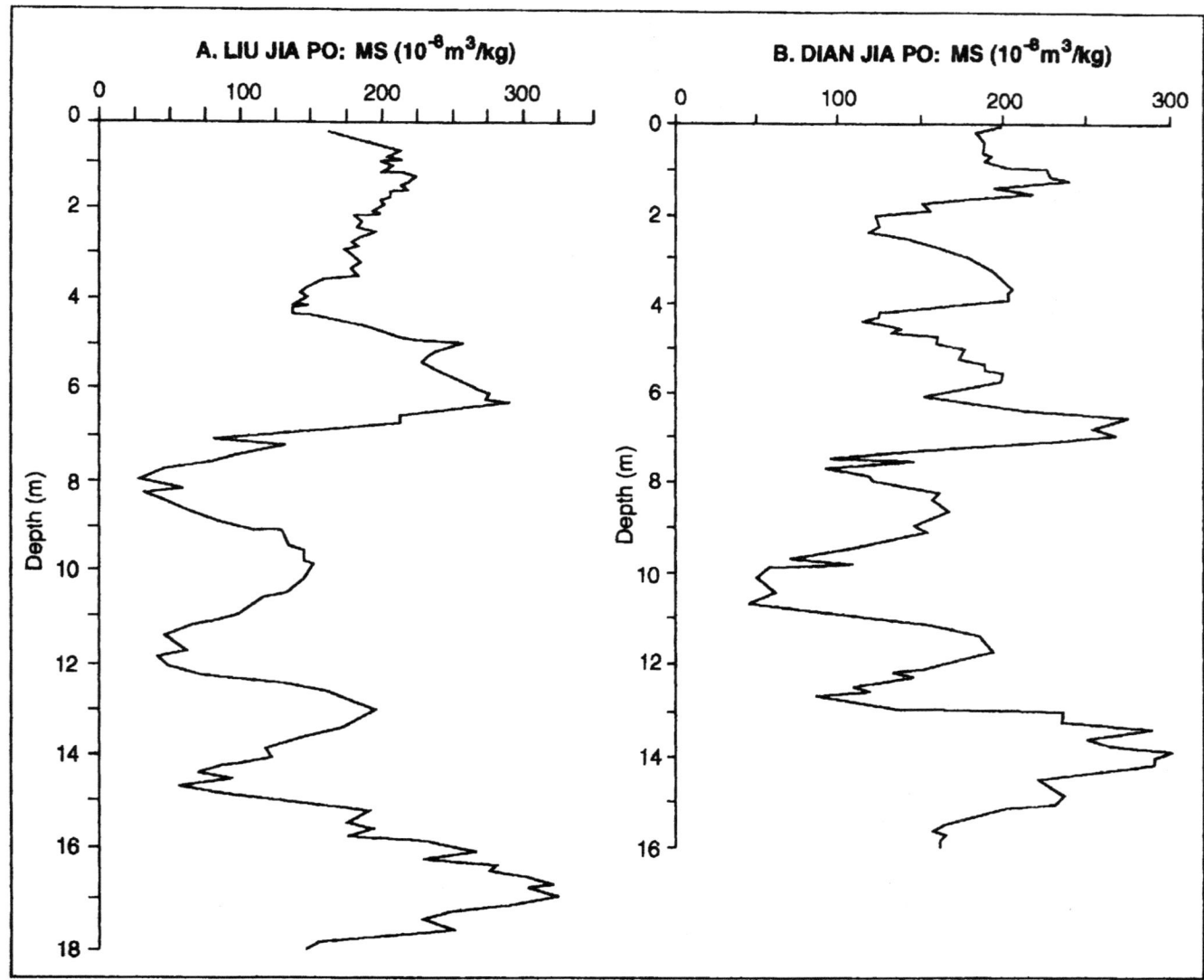

Figure 10. Magnetic susceptibility curves for (a) Liu Jia Po, and (b) Duan Jia Po.

the pattern of carbonate enrichment (Fig. 11c). Depth functions of carbonate (0-17%) are less clear than those observed in the Lanzhou sections (Kemp *et al.* 1995) because the greater amounts of leaching and lower rates of loess deposition have led to welding of pedogenic profiles under the moister climates of past and present. Micromorphological analysis suggests that the carbonate peak between 3.8 and 4.8m at Liu Jia Po has resulted from the accumulation of reprecipitated calcite leached from the soil unit whose presence is indicated by both the magnetic susceptibility and the median grain curves between 0.8 and 2.5m. The carbonate peak between depths of 1.5 and 2.5m has no associated magnetic susceptibility peak above it, suggesting that the original decalcified soil horizons have been eroded. It is concluded that the upper magnetic susceptibility peak at Liu Jia Po and the penultimate peak at Duan Jia Po mark the S_m palaeosol. It is also concluded that the S_0 palaeosol has been removed from Liu Jia Po.

Further data were then sought to test this interpretation and, specifically, to determine the number and age of any erosional intervals and estimate the amount lost as a result of erosion. Twenty block samples, approximately 10 x 5cm² in

size and with a vertical spacing of no more than 50cm, were obtained from the uppermost 5m of the section at Liu Jia Po. All samples were dated, using both thermoluminescence (TL) and infra-red stimulated luminescence (IRSL) techniques (Musson *et al.* 1994). The sampling line chosen for the luminescence work was situated 20m to the west of the principal line selected for magnetic susceptibility measurement. The lowermost date (55.1±3.4 ka) lies at a depth equivalent on our magnetic susceptibility curve of 4.7m, *i.e.* it lies below the interstadial (S_m) palaeosol and on the upper limb of the susceptibility peak indicating the S_1 soil (Fig. 11). The spacing of the dates is consistent with a fairly constant rate of loess deposition from 51 ka to 22 ka of about 120cm/ka (Musson *et al.* 1994). However, the fact that the age of the uppermost sample, only 0.5m below the present land surface, is 15.8±0.5 ka, clearly shows that the surface has been severely degraded in this area. Thus the luminescence dates confirmed the impression gained from the micromorphological work that erosion has removed all but the secondary carbonate horizon of the Mid-Holocene soil (S_o). By analogy with Duan Jia Po, and using comparative measurement of the vertical displacement typical of the

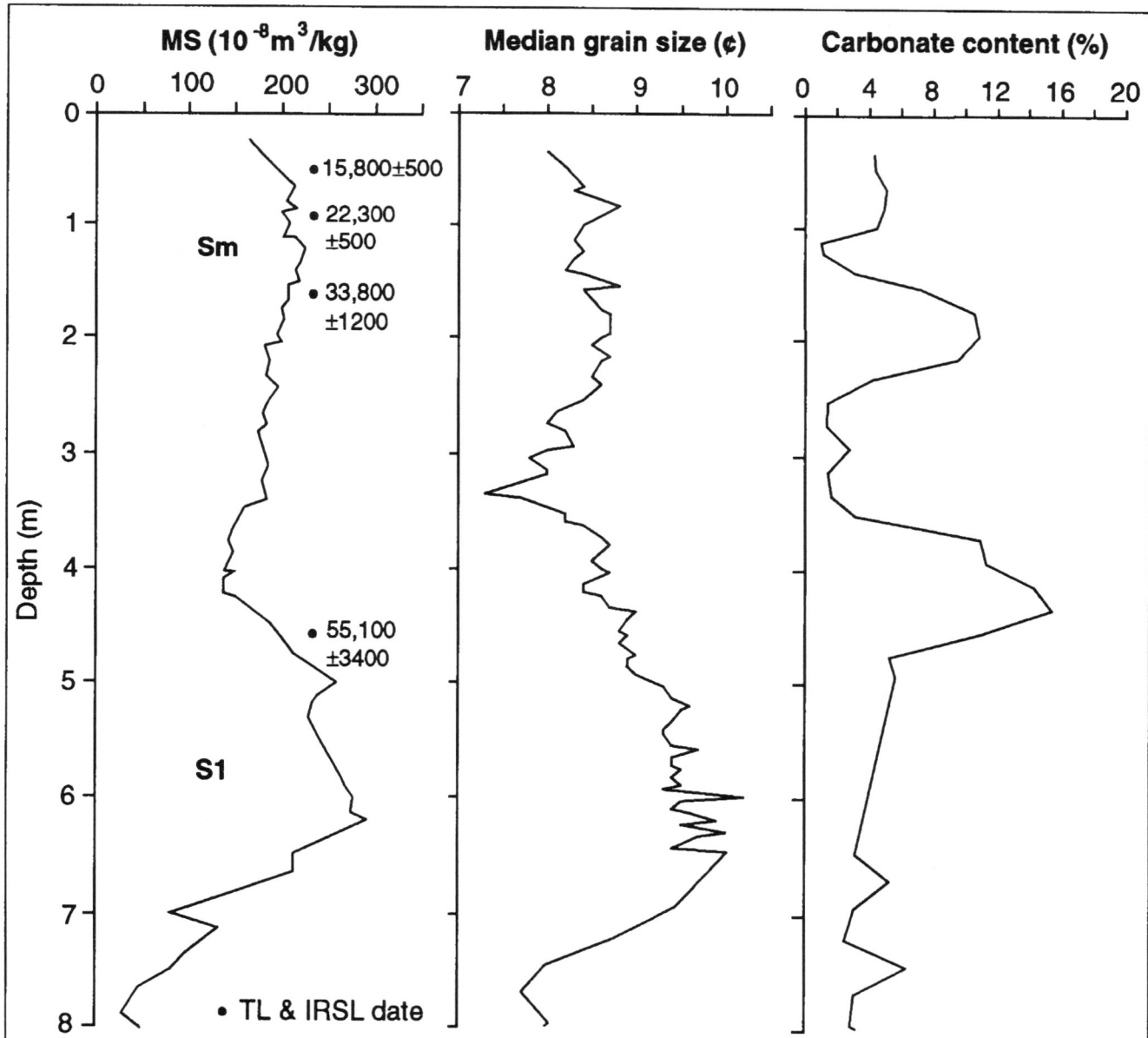

Figure 11. (a) Magnetic susceptibility, (b) median grain size, and (c) percentage carbonate content in the upper 8m of the Liu Jia Po section. Selected luminescence dates are from Musson *et al.* (1994).

carbonate-rich horizons at this site, a lowering of the surface of the loess plateau here of the order of at least 1m and perhaps as much as 3m is suggested. In the absence of evidence of concentrated erosion, such as gully systems, such severe degradation by natural agencies seems unlikely. On the other hand, this part of the plateau contains a considerable number of conical constructions which mark burial places of the nobility. These monuments, typically between 20 and 30m in height, date from the Tang Dynasty (618-907 A.D.). It seems that they were constructed by scraping off the Holocene, and some of the Late Pleistocene loess from the plateau in a manner so uniform that no morphological evidence of quarrying remains. The lesson of Liu Jia Po is that a stratigraphical method which depends upon counting down from the top is dangerous even in geomorphologically-stable sites because the degree of human interference may be severe but, at the same time, difficult to discern.

Discussion and Conclusions

The past twenty years has witnessed the rapid accumulation of evidence indicating that the loess-palaeosol sequences of the Loess Plateau of North China contain a detailed record of cold-dry/warm-moist climatic variations during the last 2.5Ma. Increasingly, interpretation of these sequences has used the palaeosols and the interbedded loess units as rather different climate proxies. On the one hand, the palaeosols are viewed as phases of relative landscape stability, with dust inputs at a minimum and pedogenetic alteration at a maximum, *i.e. as proxies of the efficacy of the summer monsoon* (e.g. An *et al.* 1993). On the other hand, the loessic units have often been interpreted in terms of their mean or median particle sizes as indicators of mean wind speeds, *i.e. as proxies of the efficacy of the winter monsoonal flow* (e.g. Rutter 1992).

The study of the high resolution profiles available on the western margins of the Loess Plateau is likely to refine this simple model. It has already been suggested, for example, that variations in winter and summer monsoonal wind systems were not in phase in the record of the section at Baxie, near Lanzhou (An *et al.* 1993). Nevertheless, it has been shown that, even here, there remains a risk of misreading local lithostratigraphy, particularly where it is poorly constrained by the paucity of datable materials, because of the instability of slopes in this mountainous, tectonically-active terrain and its violent rainfall regime. In order to make the most of the quality of the palaeoclimatic record offered by these high resolution sites it will be necessary, at the least, to use multiple adjacent sections and, ideally, to excavate series of well sections. Even this approach is not without its problems, however, as the case of Gaolanshan has shown.

It is concluded that the loess-palaeosol sequences of North China are pedosedimentary systems, the detailed study of which requires the complementary use of a broad range of field and laboratory techniques. These sequences, even where of century to decadal resolution, vary laterally to a notable degree as a result of different degrees of erosion, both natural and human-induced, superimposition or 'welding' of one palaeosol upon another, fine variations in the balance between incoming loessic dust and pedogenetic modification, and catenary changes in the palaeosols. The very high resolution in the loess record in the north-west of the Loess Plateau is a product of high sedimentation rates and so of a highly dynamic aeolian regime. The possibility must be countenanced, therefore, that such enhanced climatic resolution may carry with it a high likelihood of disparities between adjacent

sections, and that such localised variations may be masked in the more deeply weathered loess of the south and east. Thus, variations of a resolution finer than the gross global climatic signals demonstrated for an increasing number of sites across the Loess Plateau may be detectable only in the north and west. The evaluation of such variations at the site scale has clear implications for inter-regional correlation of the loess series of the western Loess Plateau with better-documented and more tightly temporally-constrained sequences in quite different climates, including the current type section for the Chinese loess stratigraphy at Luochuan, Shaanxi Province, over 500km to the east of Lanzhou.

Acknowledgements

The work reported here formed part of a programme funded by NERC Grant GST/02/540. This could not have been completed without several years of commitment and consistent material support by the late Professor Wang Jingtai of the Gansu Academy of Sciences at Lanzhou. Thanks are due to the Commission of the European Communities (Contract CI1*-CT92-0004) and the Government of Gansu Province for financing the excavation of the well sections on both Gaolanshan and Jiuzhoutai. Field assistance by the Xian Laboratory for Loess Research and Quaternary Geology (Academia Sinica) is also acknowledged. We thank A.G. Wintle and F.M. Musson for the luminescence dating results for Liu Jia Po, and A.G. Wintle for improving an earlier draft of this paper.

References

AN, Z.S., WU, X.H., WANG, P.X., WANG, S.M., DONG, G.R., SUN, X.J., ZHANG, D.E., LU, Y.C., ZHENG, S.H. & ZHAO, S.G. (1991a). Changes in the monsoon and associated environmental changes in China since the last interglacial. *In:* Liu, T.S. (ed.) *Loess, Environment and Global Change.* Science Press, Beijing, 1-29.

AN, Z.S., KUKLA, G.J., PORTER, S.C. & XIAO, J.L. (1991b). Magnetic susceptibility evidence of monsoon variation on the Loess Plateau of central China during the last 130,000 years. *Quaternary Research*, 36: 29-36.

AN, Z.S., PORTER, S.C., ZHOU, W.J., LU, Y.C., DONAHUE, D.J., HEAD, M.J., WU, X.H., REN, J.Z. & ZHENG H.B. (1993). Episode of strengthened summer monsoon climate of Younger Dryas age on the Loess Plateau of central China. *Quaternary Research*, 39: 45-54.

BURBANK, D.W. & LI, J.J. (1985). Age and palaeoclimatic significance of the loess of Lanzhou, north China. *Nature* 316: 429-431.

CHEN, F.H. LI, J.J. & ZHANG, W.X. (1991). Loess stratigraphy of the Lanzhou profile and its comparison with deep-sea sediment and ice core record. *GeoJournal*, 24: 201-209.

DERBYSHIRE, E. (1984). Granulometry and fabric of the loess at Jiuzhoutai, Lanzhou, People's republic of China. *In:* Pecsi, M. (ed.) *Lithology and Stratigraphy of Loess and Paleosols.* Hungarian Academy of Sciences, Budapest, 97-103.

DERBYSHIRE, E., WANG, J.T., JIN, Z.X., BILLARD, A., EGELS,

Y., KASSER, M., JONES, D.K.C., MUXART, T., & OWEN, L. (1991). Landslides in the Gansu loess of China. *Catena* Supplement 20, 119-145.

DIJKSTRA, T.A., DERBYSHIRE, E. & MENG, X.M. (1993). Neotectonics and mass movements in the loess of north-central China. *Quaternary Proceedings* No. 3, 93-110.

DING, Z.L., RUTTER, N., HAN, J.T. & LIU, T.S. (1992). A coupled environmental system formed at about 2.5 Ma in East Asia. *Palaeogeography, Palaeoclimatology, Palaeoecology* 94: 223-242.

HELLER, F. & LIU, T.S (1982). Magnetostratigraphic dating of loess deposits in China. *Nature* 300: 431-433.

HELLER, F. & LIU, T.S (1984). Magnetism of Chinese loess deposits. *Journal of the Royal Astronomical Society* 77: 125-141.

HELLER, F. & LIU, T.S. (1986). Palaeoclimatic and sedimentary history from magnetic susceptibility of loess in China. *Geophysical Research Letters* 13: 1169-1172.

HELLER, F., LIU, X.M., LIU, T.S., & XU, T. (1991). Magnetic susceptibility of loess in China. *Earth and Planetary Science Letters* 103: 301-310.

IMBRIE, J., HAYS, J.D., MARTINSON, D.G., McINTYRE, A., MIX, A., MORLEY, J.J., PISIAS, N.G., PRELL, W. & SHACKLETON, N.J. (1984). The orbital theory of Pleistocene climate: support from a revised chronology of the marine $\delta^{18}O$ record. *In:* Berger, A.L., Imbrie, J., Hays, J., Kukla, G., and Saltzman, B. (eds.) *Milankovitch and Climate.* D. Reidel, Hingham, Massachusetts, Part 1: 269-305.

KEMP, R.A., DERBYSHIRE, E., MENG, X.M., CHEN, F.H. & PAN, B.T. (1995). Pedosedimentary reconstruction of a thick loess-paleosol sequence near Lanzhou in north-central China. *Quaternary Research*, 43: 30-45.

KUKLA, G. (1987). Loess stratigraphy in central China. *Quaternary Science Reviews* 6: 191-219.

KUKLA, G. & AN, Z.S. (1989). Loess stratigraphy in central China. *Palaeogeography, Palaeoclimatology, Palaeoecology* 72: 203-225.

KUKLA, G., HELLER, F., LIU, X.M., XU, T.C., LIU, T.S., & AN, Z.S. (1988). Pleistocene climates in China dated by magnetic susceptibility. *Geology* 16: 811-814.

KUKLA, G., AN, Z.S., MELICE, J.L., GAVIN, J., & XIAO, J.L. (1990). Magnetic susceptibility of Chinese loess. *Transactions of the Royal Society of Edinburgh: Earth Sciences* 81: 263-288.

LI, J.J., ZHU, J.J., KANG, J.C., CHEN, F.H., FANG, X.M., MU, D.F., CAO, J.X., TANG, L.Y., ZHANG, Y.T. & PAN, B.T. (1992). The comparison of Lanzhou loess profile with Vostok ice core in Antarctica over the last glaciation cycle. *Science in China* (Series B) 35: 476-487.

LI, J.J. & FENG, Z.D. (1988). Late Quaternary monsoon patterns on the Loess Plateau of China. Earth Surface Processes and Landforms, 143: 125-135.

LIKHAREV, I.M. & RAMMEL'MEIER, E.S. (1952). *Terrestrial mollusks of the Fauna of the USSR.* Israel Program for Scientific Translation (1962).

LIU, T.S. (1958). Preliminary investigation on the loess of Shanxi and Shaanxi Provinces in the middle reaches of the Huanghe River. *Quaternaria Sinica,* 1: 255-257 (in Chinese).

LIU, T.S. (ed.) (1964). *Loess on the middle reaches of the Yellow River.* Science Press, Beijing, 234pp. (in Chinese).

LIU, T.S. (ed.) (1985). *Loess and the Environment.* Science Press, Beijing. 215pp.

LIU, T.S., AN, Z.S., YUAN, B.Y., & HAN, J.M. (1985). The loess paleosol sequence in China and climatic history. *Episodes* 8: 21-28.

LIU, T.S. & CHANG, T.H. (1964). The'huangtu' (loess) of China. *Report of the Sixth INQUA Congress,* Warsaw 1961, 4: 503-524.

LIU, T.S. & DING, Z.L. (1993). Stepwise coupling of monsoon circulations to global ice volume variations during the late Cenozoic. *Global and Planetary Change,* 7: 119-130.

LIU, T.S., DING, Z.L., CHEN, M.Y. & AN, Z.S. (1989). The global surface energy system and the geological role of wind stress. *Quaternary International* 2: 43-54.

LIU, T.S. & WANG, K.L. (1964). Some aspects on the Quaternary stratigraphy in North China. *In:* Liu, T.S. (ed.) *On the Quaternary Geology in China.* Science Press, Beijing, 65-76 (in Chinese).

LIU, T.S. & YUAN, B.Y. (1982). Quaternary climatic fluctuation - a correlation of records in loess with that of the deep sea core V28-238. *In:* Research on Geology-1. Institute of Geology, Academia Sinica (IGAS), Cultural Relics Publishing House, Beijing, 113-121 (in Chinese).

LIU, T.S. & YUAN, B.Y. (1987). Paleoclimatic cycles in Northern China (Luochuan loess section and its environmental implications). *In:* Liu, T.S. (ed.) *Aspects of Loess Research.* China Ocean Press, beijing, 3-26.

LIU, X.M., ROLPH, T.C., BLOEMENDAL, J, SHAW, J. & LIU, T.S. (1994). Remanence characteristics of different magnetic grain size categories at Xifeng, central Chinese Loess Plateau. *Quaternary Research* 42: 162-165.

MUSSON, F.M., CLARKE, M.L. & WINTLE, A.G. (1993). Luminescence dating of loess from the Liujiapo section, central China. *Quaternary Science Reviews* 13: 407-410.

MUXART, T., BILLARD, A., WANG, J.T. & DERBYSHIRE, E. (1994). Variation in runoff on steep, unstable slopes near Lanzhou, China: initial results using rainfall simulation. *In:* Kirkby, M.J. (ed.) *Process Models and Theoretical Geomorphology.* John Wiley, Chichester, 337-355.

PORTER, S.C., AN, Z.S. & ZHENG, H.B. (1992). Cyclic Quaternary alluviation and terracing in a nonglaciated drainage basin on the north flank of the Qinling Shan, central China. *Quaternary Research*, 38: 157-169.

ROLPH, T.C., SHAW, J., DERBYSHIRE, E., & WANG, J.T. (1989). A detailed geomagnetic record from Chinese loess. *Physics of the Earth and Planetary Interiors*, 56: 151-164.

ROLPH, T.C., SHAW, J., DERBYSHIRE, E. & WANG, J.T. (1993). The magnetic mineralogy of a loess section near Lanzhou, China. *In:* Pye, K. (ed.) *The Dynamics and Environmental Context of Aeolian Sedimentary Systems*, The Geological Society of London, 311-323.

SPARKS, B.W. (1961). The ecological interpretation of Quaternary non-marine Mollusca. *Proceedings of the Linnean Society of London*, 172: 71-80.

RUTTER, N. (1992). Presidential Address, XIII INQUA Congress 1991: Chinese loess and Global Change. *Quaternary Science Reviews*, 11: 275-281.

WANG, Y.Y. (ed.) (1982). *Loess and Quaternary Geology.* People's Press, Shaanxi, 20-47.

ZHANG, H.C. (1992). Climatic change and eolian sedimentation and geomorphological processes of loess. *Lanzhou University Journal*, Supplement 1992: 21-28.

ZHANG, J. & LIN, Z.G. (1992). *Climate of China.* John Wiley & Sons Ltd., Chichester, 376pp.

ZHENG, H.B., AN, Z.S., SHAW, J. & LIU, T.S. (1991). A detailed terrestrial geomagnetic record for the interval 0 - 5.0 Ma. *In:* Liu, T.S. (ed.) *Loess, Environment and Global Change.* Science Press, Beijing, 147-156.

ZHOU, L.P., OLDFIELD, F., WINTLE, A.G., ROBINSON, S.G. & WANG, J.T. (1990). Partly pedogenic origin of magnetic variations in Chinese loess. *Nature*, 346: 737-739.

Quaternary Proceedings No. 4, 1995 19-26
© Quaternary Research Association, Cambridge.

A Comparison of Magnetic Fabrics from Loessic Silts Across the Tibetan Front, Western China.

M.L. Clarke

M.L. Clarke, 1995 A comparison of magnetic fabrics from loessic silts across the Tibetan front, western China, In *Wind Blown Sediments in the Quaternary Record* (Edward Derbyshire). Quaternary Proceedings No. 4, John Wiley & Sons Ltd., Chichester, pp. 19-26.

Abstract

Anisotropy of magnetic susceptibility has been measured in loess and palaeosols from areas across the Tibetan Front, China. A range of fabrics from aeolian to fluvial have been observed in these sediments. Loess from Labrang, in the mountains which form the northeastern edge of the Tibetan Plateau, showed a randomly-orientated fabric indicative of an airfall deposit. Loesses and palaeosols from the western Loess Plateau, around the city of Lanzhou, showed both aeolian fabrics with preferential orientation of grain long axes and slope-induced fabrics. A control sample of horizontally-bedded loessic alluvium showed a strongly anisotropic fabric dominated by a high degree of foliation, about six times greater than the loess deposits, which were all weakly foliated.

KEYWORDS: loess, fabric, China, palaeoclimate.

M.L. Clarke, Institute of Earth Studies, University of Wales, Aberystwyth, Dyfed SY23 3DB, Wales.

Introduction

The thick loess and palaeosol sequences of central China are thought to represent a comprehensive proxy record of Quaternary palaeoclimatic fluctuation. Changes in magnetic susceptibility throughout these profiles has been linked to the deep sea oxygen isotope records (Kukla *et al.* 1988; Hovan *et al.* 1989) and the Vostok ice core record (Chen Fahu *et al.* 1991). However, these comparisons rely on the premise that the loess profiles are free from unconformities, neotectonic-related mass movement or other reworking since deposition. The aim of this study was to test the assumption that the loess of China and the Tibetan Grasslands is a true aeolian deposit and thus valid as a proxy palaeoclimate indicator.

Anisotropy of magnetic susceptibility may be used to establish the depositional environment of the magnetic grains found within a sediment matrix, and thus of the sediment as a whole. The orientation of magnetic grains within a sediment is affected by gravity, the topography of the depositional surface and the aligning forces of transporting currents (Hamilton & Rees 1970; Hrouda 1982). The orientation of grain long axes depends upon the dominant force operating on the grains. On horizontal surfaces, there are no dominant forces acting on the grains, thus platy (oblate) grains will lie with their long axes parallel to the bedding plane and cigar-shaped (prolate) grains will have a random orientation. Thus the sediment fabric is entirely foliated parallel to the bedding plane, with no superimposed lineation. However, slope fabrics reflect the dominant gravitational forces on the sediment, with fabrics still dominated by foliation parallel to the bedding plane but there is a superimposed lineation as prolate grain long axes roll

perpendicular to the slope. The degree of lineation increases with increasing slope angle. Weak transporting currents orientate prolate grain long axes parallel to the direction of current flow, but the degree of imbrication is sufficiently small to allow the majority of oblate grains to remain parallel to the bedding direction and thus the fabric is dominantly foliate (Hrouda 1982; Hrouda & Tarling 1993). However, as the current increases, the prolate grains roll and their long axes become aligned transverse to the current direction whereas the oblate grains become tilted 5-20° away from the bedding plane; this leads to superimposed lineations upon the oblate fabric (Hrouda & Tarling 1993). As loess is primarily an air fall deposit, its grains are affected mainly by gravity, but may also be orientated by wind currents or by water in the form of rainsplash or slurry flow.

Preferred grain orientations indicative of past wind directions have been found in the Vicksberg Loess, Mississippi (Matalucci *et al.* 1969) and in loess at Liujiapo, just east of the city of Xian (Thistlewood & Sun 1991). Thistlewood and Sun (1991) found a weak sub-horizontal fabric with a preferred WNW-ESE orientation which was thought to coincide with the sediment transport direction. Also in China, Derbyshire *et al.* (1988) used both magnetic and optical studies on loess from Jiuzhoutai, concluding that the primary aeolian fabric is isotropic and that other fabrics result from reworking by slope and alluvial processes. Magnetic measurements on alluvially redeposited loess from the central part of the Loess Plateau have shown that, compared with windblown loess, alluvial loess shows a sub-horizontal fabric with a strongly developed foliation and a strong correlation between foliation and degree of anisotropy (Liu *et al.* 1988).

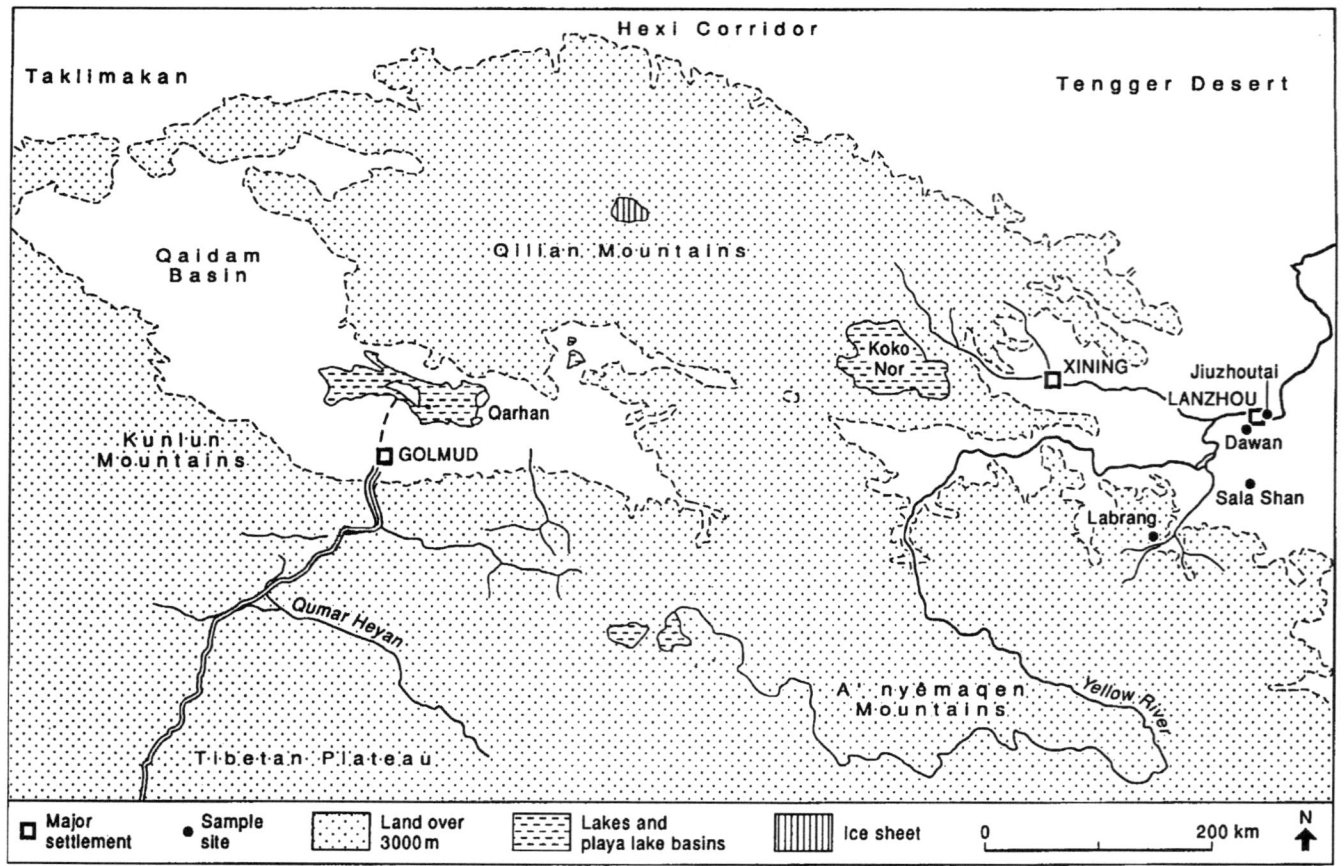

Figure 1. Map of the Tibetan Front showing sample localities.

The Tibetan Front and Sample Sites

There is a steep climatic gradient existing across the Chinese Loess Plateau, from the middle reaches of the Yellow River, an area affected by the summer monsoon (a combination of the East Asian monsoon and the southwest Indian monsoon), to the arid western fringes of the Loess Plateau (around Lanzhou) and the hyper-arid northwestern deserts of Qinghai and Xinjiang, where the climate is dominated by the Mongolian-Siberian anticyclone. The Tibetan Front is the area of northwestern China and Tibet consisting of high mountains and arid basins lying in the rainshadow of the Tibetan Plateau (Fig. 1).

Four sites within this area were chosen for fabric studies. Two of the sites, Jiuzhoutai (36° 00'N, 103°45'E) and Dawan (35°54'N, 103°12'E), have thick (>290 metre) loess-palaeosol sequences resting on terrace gravels of the Yellow River. These profiles are located close to the city of Lanzhou, and are approximately 75km apart. Loess was sampled from short sections in the upper part of both sections corresponding to Late and Middle Pleistocene, Malan and Lishi Loesses; intervening palaeosols were also sampled. Figure 2 shows the changes in mass specific low field susceptibility (χ_{LF}; measured on a Bartington MS1 susceptibility bridge) through the two profiles. The samples from the Scorpion Pit, Jiuzhoutai came from depths of 47 to 53 metres below the top of loess succession. The Scorpion Pit is situated on the south side of Jiuzhoutai on a slope of 35° falling in an easterly direction (110°) to the Yellow River. The samples from Dawan were taken from the east side of the Shua Jia Valley at depths of 20 to 35 metres from the top of the loess section. These were excavated from five different sample pits, each of which had a different orientation to act as a control on systematic sampling

error. The slope at Dawan is approximately 35° WNW (290°) down to the Shua Jia River.

Loess of unknown age was sampled from a previously undescribed site at Labrang (35°20'N, 102°50'E), where a 60 metre thick loess terrace occurs on the north side of the Daxia River, south of the Labrang Tibetan Monastery (Fig. 3). This site is located outside the Loess Plateau in a mountainous area known as the Tibetan Grasslands, which lies at an altitude of over 3000 metres and forms the eastern fringes of the Tibetan pastureland. The Labrang loess has a different source from the Loess Plateau sediment and was probably formed within the local mountain environment (Clarke 1992). A control sample of visibly stratified loessic alluvium (Fig. 4) was sampled from a Ba Xie River cutting at Sala Shan (35°50'N, 103° 00'E), which is situated east of the town of Linxia and 60km SSW of Lanzhou.

All of the samples described here were taken from consolidated loess by the following sampling procedure. The loess face was cleaned and then excavated to form blocks approximately 15cm across and 35cm in depth. The top surface was levelled to horizontal using a small 2 way-spirit level and a north arrow was marked on the level surface before the block was removed from the face. These orientated blocks were then sub-sampled into cubes and placed in 2.2cm³ perspex boxes; thus the block was cut into approximately 14 cubes, representing every 2.5cm down the section. At least 2 cubes were prepared from each sample height, making a total of at least 28 cubes from each block.

Anisotropy of Magnetic Susceptibility

Magnetic fabric is usually defined in terms of the principal

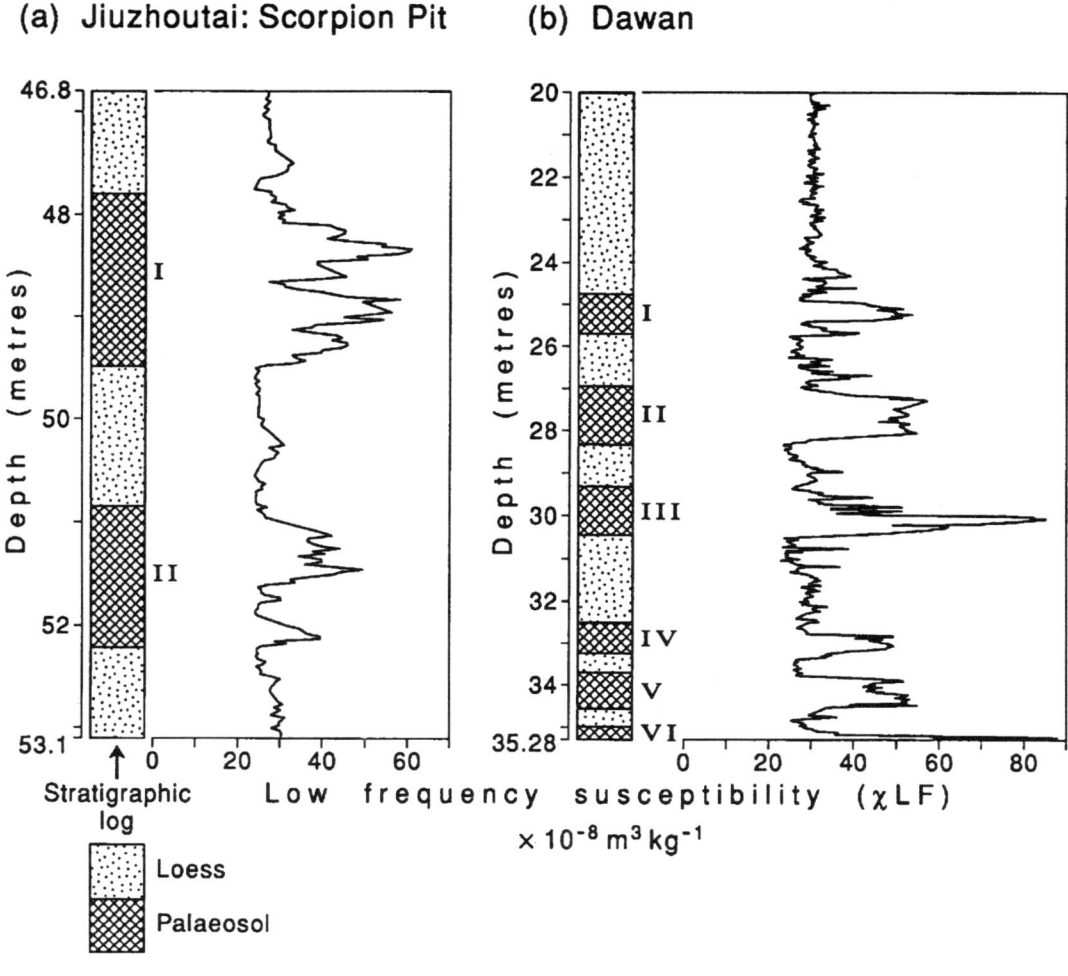

Figure 2. Mass specific low field susceptibility (χLF) plots of the loess and palaeosol sequences at (a) Jiuzhoutai and (b) Dawan.

Figure 3. View southeastward up the Daxia River showing the 60 metre thick loess terrace at Labrang, with a factory chimney in the upper centre of the photograph for scale.

Figure 4. Stratified loessic alluvium of the Ba Xie River from Sala Shan. The knife indicates the sample location.

components of the susceptibility ellipsoid, which is the summation of the susceptibility of all individual magnetic grains. Kmax is defined as the axis of maximum susceptibility, Kint is the intermediate and Kmin is the minimum susceptibility axis, where K is volume susceptibility. If all three axes were equal (Kmax=Kint=Kmin) the susceptibility ellipsoid would correspond to an average spherical shape for all the grains in the sample, and the sample would be isotropic; the net shape of the anisotropy ellipsoid strongly reflects the shape of the magnetic grains within a sediment (Tarling & Hrouda 1993). An anisotropic fabric would suggest the relative absence of spherical grains and the presence of predominantly oblate or prolate grains. Prolate grains are cylindrical or cigar-shaped and characterised by Kmax > Kint $\cong$ Kmin. A dominance of prolate grains gives a linear fabric, where lineation is defined as: l = (Kmax-Kint)/K. Oblate grains are disc or pancake-shaped and are characterised by Kmax $\cong$ Kint >Kmin. A dominance of oblate grains gives a foliated fabric, where foliation is defined as f = (Kint - Kmin)/K. The total anisotropy (H) gives a measure of the strength of the fabric, where H = (Kmax - Kmin)/K. The parameter q (the azimuthal anisotropy quotient) has also been used, where q = (Kmax - Kint)/[(Kmax + Kint)/2 - Kmin]. It reflects the relative importance of lineation and foliation within the fabric, varying from 0 in pure foliated fabrics to 2 in purely linear fabrics, the change from foliation to lineation occurring at 0.67 (Hamilton & Rees 1970).

Measurement of Magnetic Fabric

Anisotropy of magnetic susceptibility was measured on a Molspin Minisep Anisotropy Delineator and bulk susceptibility on a Minisep Bridge. 2.2cm³ perspex containers holding the

orientated loess samples were spun at a frequency of 6Hz about a vertical axis within two sets of orthogonal Helmholtz coils. One set of coils produces a field in the order of 0.7mT, and the other set detects a field produced by the sample, which varies with twice the rotation frequency. The magnitude of the detected field is proportional to the sample anisotropy of magnetic susceptibility about the axis of rotation. The sample is rotated about three orthogonal axes and the intensity, declination and inclination of the magnetic susceptibility are measured in each of the principal axes to give the shape of the susceptibility ellipsoid. The data are presented here using filled circles to represent the maximum axes, open circles to represent the intermediate axes and triangles to represent the minimum axes, plotted on equal area lower hemisphere projections.

Results

The 399 sub-samples measured from the Scorpion Pit at Jiuzhoutai were arranged into stratigraphic units (loess I to loess III) on the basis of their low field mass-specific susceptibility (Table 1). Palaeosol I and II are believed to correspond to the last interglacial palaeosols (Chen Fahu et al. 1991). The fabric of the loess samples shows a weak foliation, with q values close to the linear-foliated boundary, ranging from 0.525 to 0.681. The palaeosol samples show weaker fabrics than those of the loess, probably due to the disrupting effects of bioturbation and weathering. Typical lineation-foliation and foliation-total anisotropy plots for the loess and palaeosols are shown in figure 5. The lower hemisphere stereographic projection of principal susceptibility axes for loess I is shown in figure 6a. The Kmax axes (equivalent to grain long axes) presented as closed circles, and the Kint axes, presented as open circles, are

Table 1. Magnetic fabric parameters for the Scorpion Pit, Jiuzhoutai, Dawan, Labrang and Sala Shan; n stands for the number of cubes measured.

Stratigraphic unit	Lineation	Foliation	Total anisotropy	q	n
JIUZHOUTAI					
Loess I	0.0154±0.0083	0.0244±0.0138	0.0387±0.0178	0.525±0.305	103
Palaeosol I	0.0139±0.0092	0.0125±0.0079	0.0264±0.0134	0.753±0.380	83
Loess II	0.0181±0.0128	0.0209±0.0132	0.0400±0.0204	0.681±0.366	78
Palaeosol II	0.0033±0.0029	0.0118±0.0089	0.0151±0.0106	0.267±0.159	32
Loess III	0.0178±0.0147	0.0193±0.0131	0.0371±0.0215	0.613±0.393	103
DAWAN					
Loess I	0.0078±0.0063	0.0188±0.0063	0.0266±0.0085	0.345±0.199	326
Palaeosol I	0.0055±0.0037	0.0087±0.0048	0.0142±0.0068	0.534±0.271	48
Loess II	0.0074±0.0043	0.0175±0.0082	0.0248±0.0108	0.386±0.252	128
Palaeosol II	0.0100±0.0053	0.0176±0.0082	0.0276±0.0088	0.472±0.273	44
Loess III	0.0187±0.0125	0.0266±0.0143	0.0453±0.0188	0.561±0.349	125
Palaeosol III	0.0041±0.0017	0.0126±0.0046	0.0167±0.0053	0.320±0.231	36
Loess IV	0 0186±0.0126	0.0301±0.0182	0.0488±0.0254	0.527±0.299	154
Palaeosol IV	0.0259±0.0119	0.0236±0.0084	0.0495±0.0156	0.706±0.255	40
Loess V	0.0337±0.0082	0.0305±0.0110	0.0642±0.0144	0.740±0.235	36
Palaeosol V	0.0143±0.0147	0.0218±0.0172	0.0361±0.0281	0.461±0.296	56
Loess VI	0.0135±0 0071	0.0378±0.0428	0.0514±0.0482	0.343±0.107	24
Palaeosol VI	0.0067±0.0020	0.0138±0.0057	0.0236±0.0074	0.338±0.115	7
LABRANG					
Terrace loess	0.0137±0.0080	0.0138±0.0102	0.0275±0.0142	0.695±0.405	54
SALA SHAN					
Loessic alluvium	0.0060±0.0045	0.0670±0.0133	0.0610±0.0121	0.093±0.472	29

both scattered around declinations of 150-355°; the apparent reversibility of these axes indicates the oblate nature of the susceptibility ellipsoid. The Kmin axes (triangles) are distributed about a declination of 40° - 70° with varying dip angles of 3 - 87°. This arrangement of axes implies that the magnetic grains in the sediment matrix are not lying on a horizontal bedding plane but are distributed on slopes of varying angle. This is borne out by the wide angle of Kmax distribution.

1024 sub-samples were measured from 12 stratigraphic units at Dawan, designated loess I to palaeosol VI. Most of the loesses (except loess V) show a weak anisotropy with mean q values ranging from 0.34 to 0.56, indicating an oblate susceptibility ellipsoid and a foliated fabric. Loess V is more strongly anisotropic with a relatively prolate susceptibility ellipsoid indicated by a mean q value of 0.74. The palaeosols from the Dawan section show weakly foliated fabrics and a lower total anisotropy than the loess in which they were developed, probably due to bioturbation. The stereographic projection in figure 6b shows the 326 samples representing the 5 metres of loess I. The principal susceptibility axes suggest a horizontal foliated fabric with a preferred grain orientation trending ENE-WSW.

The Labrang loess is weakly anisotropic and equally linear and foliated, illustrated by a mean q value of 0.695. The stereographic projection (Fig. 6c) shows no preferred grain orientation. The loessic alluvium from Sala Shan is strongly anisotropic and dominated by a well-developed foliation, six times greater than any of the other samples studied (Fig. 5). The

q value of 0.093 indicates the highly oblate nature of the susceptibility ellipsoid. The lower hemisphere stereographic projection (figure 6d) shows close clustering of Kmin axes and a variable declination in Kint and Kmax axes, as would be expected by deposition within shallow braided channels affected by sinusoidal changes in current direction.

Discussion

The control sample of redeposited loessic alluvium from Sala Shan shows a well developed foliation and strong anisotropy, in agreement with the findings of Liu et al. (1988). The other samples tested in this experiment show much weaker foliation indicating that their fabrics are not the result of alluvial processes.

The Scorpion Pit at Jiuzhoutai is excavated into a steep (35°) eastward facing (110°) slope of the mountain overlooking the Yellow River. The palaeotopography of the site (and that at Dawan) is hard to gauge due to the difficulty in identifying palaeosols in this region due to its arid climate. Kmin axes of the Jiuzhoutai samples suggest a northeasterly palaeo-slope direction with the low total anisotropy and foliation (Table I; compared to the alluvial fabric from Sala Shan) indicating that current flow was not a dominant factor in deposition. Thermoluminescence dates from the Scorpion Pit range from 123.2±13.0 under soil II to 81.0±5.9ka above the upper soil (I) and suggest that the loess is Middle and Late Pleistocene in age, with the dual palaeosol representing the last interglacial (Chen

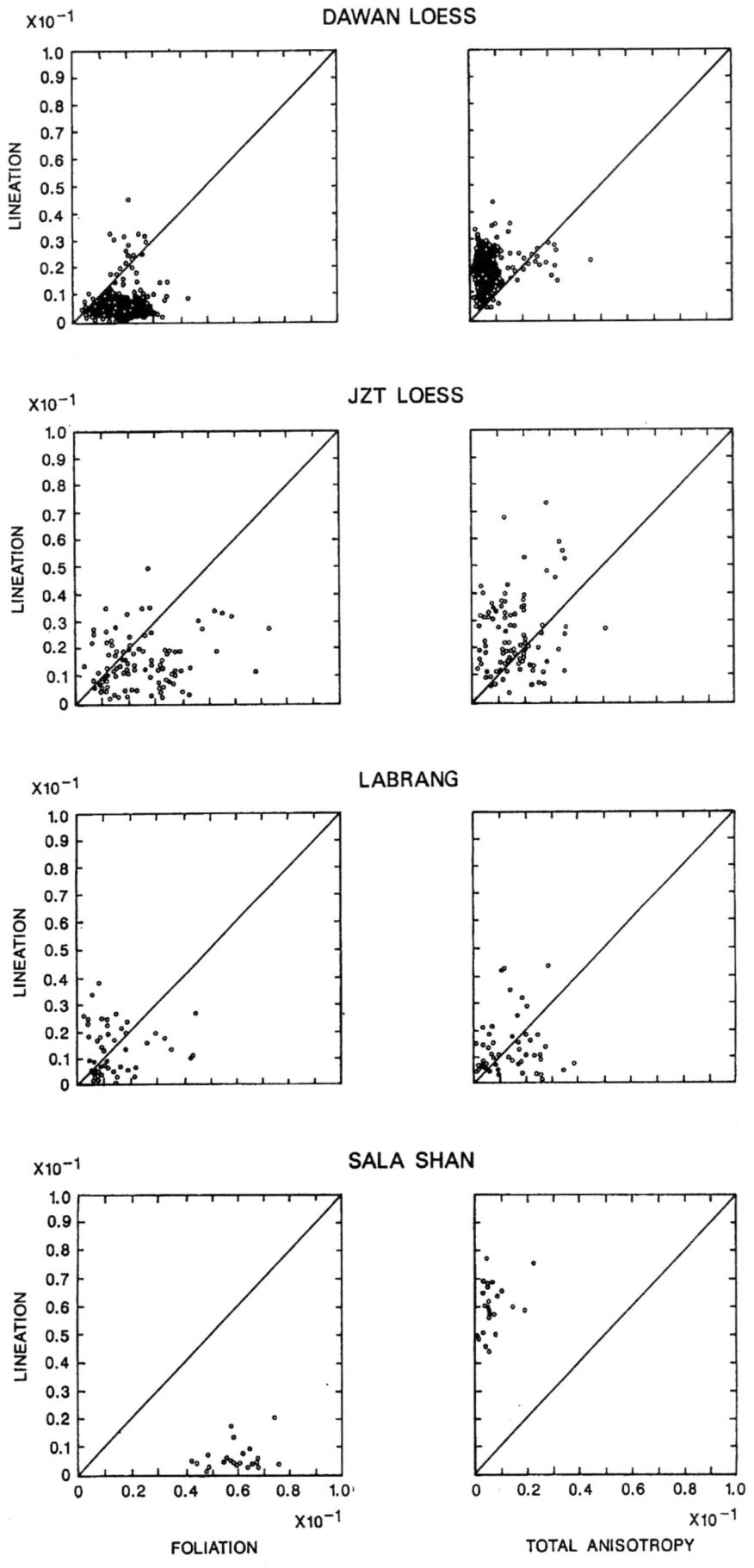

Figure 5. Typical lineation-foliation and foliation-total anisotropy plots for loess from the four sites.

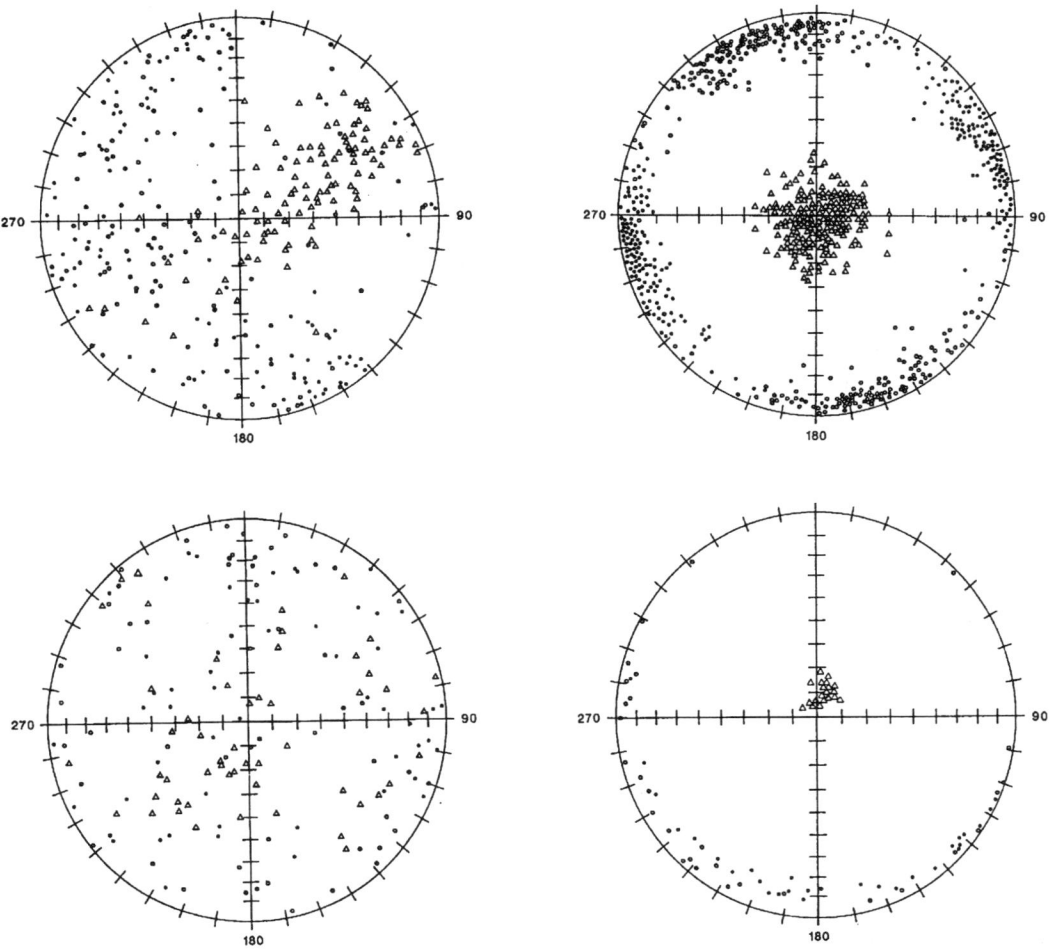

Figure 6. Lower hemisphere stereographic projection of principal susceptibility axes for (a) Jiuzhoutai loess; (b) Dawan loess; (c) Labrang loess; (d) Sala Shan loessic alluvium.

Fahu *et al.* 1991). The palaeo-slope was thus a Pleistocene surface, with a northeasterly (~55° NE) downslope orientation, indicated by the lower hemisphere stereographic projection, compared to the present easterly slope orientation of 110° ESE. Thus, the current slope is an erosional one. The preferred ENE-WSW grain orientation shown by the Dawan loess does not show the high degree of foliation and anisotropy associated with alluvial deposition and thus may reflect orientation by wind. However, this is difficult to equate with the dominant climate systems of the area. Northwesterly winds associated with the Mongolian-Siberian anticyclone approach Lanzhou from the northwest via the Hexi Corridor and from the west via the Qaidam Basin and Yellow River Valley. Moisture laden winds associated with the monsoon approach Lanzhou from the southeast, enhancing the dominant NW-SE sediment transport pathway. Labrang lies to the southwest of Lanzhou, but the loess at Labrang has a different rare earth element signature to that of Lanzhou and therefore a different source (Clarke 1992; 1995). The different orientation of the Dawan sample pits shows that the preferred grain orientation is not due to a systematic sampling error. There are three possible explanations for this wind-controlled fabric orientation: a localised topographically controlled ENE-WSW wind; a northeasterly wind emanating from the Ordos Desert, in the big bend of the Yellow River; or the northwesterly Mongolian-Siberian anticyclone winds, which are sufficiently strong that

the grains lie transverse to the current direction.

The loess samples from the sixty metre thick Labrang loess terrace, which lies within the upper Daxia River Valley, shows no preferential grain orientation and a weak foliation. This random fabric is most likely to represent an air fall deposit which has not been affected by current flow, with q values indicating an equal distribution of oblate and prolate magnetic grains. The loess at this site is most likely to have been derived from cold weathering fragments or glacial outwash within the local mountain environment (Clarke 1992).

Conclusions

Anisotropy of magnetic susceptibility measurements have shown that a variety of sediment fabrics exist in loessic silts from regions of the Tibetan Front. Two aeolian fabrics have been found: at Labrang the aeolian fabric is randomly-orientated and weakly anisotropic, whereas at Dawan it is weakly foliated, showing preferential wind-controlled ENE-WSW grain orientation. Thus the parts of the Labrang and Dawan sections which were measured consist of aeolian loess which have the potential to be useful indicators of palaeoclimate. Fluvially-reworked loess from Sala Shan shows a highly oblate susceptibility ellipsoid with a strongly developed foliation and total anisotropy; this agrees with the findings of Liu *et al.* (1988) for redeposited alluvial loess from Xifeng. Middle and Late

Pleistocene loess from the Scorpion Pit, Jiuzhoutai shows a slope-induced fabric with a different slope orientation to that of the present surface and care should be taken when using this site for palaeoclimatic interpretation.

Acknowledgements

This work was undertaken as part of NERC studentship GT4/88/GS/61. I would like to thank Prof E. Derbyshire, Dr. J. Shaw and Dr. E.A. Hailwood for advice and Dr. T.C. Rolph and Dr. G. Sherwood for technical assistance. Field support was given by Dr. C.H. Scott, Dr. J.A. King, Meng Xingming and Ma Jinhui. This work would not have been possible without the enthusiasm and support of the late Prof. Wang Jingtai of the Geological Hazards Research Institute, Gansu Academy of Sciences. This is publication number 379 of the Institute of Earth Studies, University of Wales, Aberystwyth.

References

CHEN FAHU, LI JIJUN & ZHANG WEIXIN (1991). Loess stratigraphy of the Lanzhou profile and its comparison with deep-sea sediment and ice core record. *Geojournal*, 24, 201-209.

CLARKE, M.L. (1992). Formation, depositional history and magnetic properties of loessic silts from the Tibetan Front, western China. unpub PhD thesis, University of Leicester.

CLARKE, M.L. (1995). Sedimentological characteristics and rare earth element fingerprinting of Tibetan silts and their relationship with the sediments of the western Chinese Loess Plateau. *Quaternary Proceedings*, this volume.

DERBYSHIRE, E., BILLARD, A., VAN VLIET-LANOE, B., LAUTRIDOU, J-P. & CREMASCHI, M. (1988). Loess and palaeoenvironment: some results of a European joint programme of research. *Journal of Quaternary Science*, 3, 147-169.

HAMILTON, N. & REES, A.I. (1970). The use of magnetic fabric in palaeocurrent estimation. In: Runcorn, S.K. (ed) Palaeogeophysics. Academic Press, London. p.445-464.

HOVAN, S.A., REA, D.K, PISIAS, N.G. & SHACKLETON, N.J. (1989). A direct link between the China loess and marine $\delta^{18}O$ records: aeolian flux to the north Pacific. *Nature*, 340, 296-298.

HROUDA, F. (1982). Magnetic anisotropy of rocks and its application to geology and geophysics. *Geophysical Survey*, 5, 37-82.

KUKLA, G., HELLER, F., LIU XIUMING, XU TONGCHUN, LIU TUNGSHENG & AN ZHISHENG (1988). Pleistocene climates in China dated by magnetic susceptibility. *Geology*, 16, 811-814.

LIU XIUMING, XU TONGCHUN & LIU TUNGSHENG (1988). The Chinese loess in Xifeng II. A study of anisotropy of magnetic susceptibility of loess from Xifeng. *Geophysical Journal*, 92, 349-353.

MATALUCCI, R.V., SHELTON, J.W. & ABDEL-HARDY, M. (1969). Grain orientation in Vicksberg loess. *Journal of Sedimentary Petrology*, 39, 969-979.

TARLING, D.H. & HROUDA, F. (1993). *The Magnetic Anisotropy of Rocks*. Chapman and Hall, London. 217pp.

THISTLEWOOD, L. & SUN JIANZHONG (1991). A palaeomagnetic and mineral magnetic study of the loess sequence at Liujiapo, Xian, China. *Journal of Quaternary Science*, 6, 13-26.

Quaternary Proceedings No. 4, 1995 27-40
© Quaternary Research Association, Cambridge.

Magnetic Property and Particle Size Variations in the Late Pleistocene and Holocene Parts of the Dadongling Loess Section near Xining, China.

Fahu Chen, Ruijin Wu, Diana Pompei and Frank Oldfield

Fahu Chen, Ruijin Wu, Diana Pompei & Frank Oldfield, 1995 Magnetic property and particle size variations in the late Pleistocene and Holocene parts of the Dadongling loess section near Xining, In *Wind Blown Sediments in the Quaternary Record* (Edward Derbyshire). Quaternary Proceedings No. 4, John Wiley & Sons Ltd., Chichester, pp. 27- 40.

Abstract

Detailed magnetic measurements and particle size analysis have been made on the upper part of the Dadongling loess section (N36° 35'; E101° 44'), near Xining, W.China. Magnetic grain size and concentration changes record varying degrees of pedogenesis superimposed on a flux of magnetic minerals forming part of the load of aerially deposited silts and fine sands. Stratigraphic transitions between rock magnetic 'régimes' are often abrupt despite the close sample interval and rapid rates of accumulation. There is no evidence that the diagenetic transformations of iron compounds associated with pedogenesis 'overprint' the magnetic signatures below them. In the loess samples the magnetic properties reflect a unimodal, relatively coarse ferrimagnetic grain size assemblage in the silts and fine sands. In the palaeosol samples, this mode is present along with a fine mode reflecting secondary ferrimagnetic grains. By using magnetic property and particle size variations together, it is possible to envisage separating the effects of *in situ* processes from those controlling the transport of aeolian material to the site. Consistent variations in magnetic properties towards the top of L_1 may reflect 'late-glacial' oscillations. Evidence for high frequency variations in the degree of 'magnetic pedogenesis' point to the possibility of resolving the record in the section at a level of detail sufficient to permit close comparison with ice core records.

KEYWORDS: palaeosol S_1, L_1 loess, palaeosol S_0, bulk magnetic properties.

Fahu Chen, Department of Geography, University of Lanzhou, Peoples Republic of China.

Ruijin Wu, Institute of Geography and Limnology, Nanjing, Peoples Republic of China.

Diana Pompei and Frank Oldfield, Department of Geography, University of Liverpool, Liverpool L69 3BX, UK.

Introduction

In the wake of demonstrations that the magnetic susceptibility record in Chinese Loess sections closely parallels the sequence of oxygen isotope stages in deep sea sediments (Liu *et al.* 1985; Kukla *et al.* 1988, 1989), many authors have begun to characterise the magnetic properties of loess in increasing detail. From this research there emerges a clear indication that the changes in magnetic properties that appear to be so responsive to major shifts in climate, reflect above all, the interplay between the effects of atmospherically deposited, detrital magnetic minerals within the loess, and the 'signature' of secondary, pedogenically derived magnetic minerals in palaeosols, (Zhou *et al.* 1990; Zheng *et al.* 1991; Maher & Thompson 1991, 1992; Banerjee *et al.* 1993; Liu *et al.* 1993). Most of the detailed evidence for this interpretation comes from the classic sections of Xifeng & Luochuan, though it is now clear from more extensive studies (*e.g.* Banerjee *et al.* 1993; Evans & Heller 1994) that the generalization holds true

over a very large area; moreover that the pedogenic components reflecting the warm stage intervals in the sequences are responsible for the main variations both spatial and temporal that we perceive in the record.

Most published magnetic studies have dealt with the record of major changes linked to orbital forcing on timescales of tens or hundreds of thousands of years. Far fewer have considered in detail the record of change *within* any given major isotope stage. In the central and eastern parts of the zone of loess accumulation, such a task is made more difficult by the relatively low resolution of the record and by the possibility of 'overprinting' by palaeosol signatures, as the effects of pedogenic processes during major soil forming intervals are superimposed onto and so transform the magnetic properties of the immediately underlying loess upon and above which the palaeosol formed.

Many of the river valley basins along the western margins of the Loess Plateau provide good conditions for dust deposition and the rates of accumulation are much higher than further east. For example, palaeosol S_1, formed during Isotope Stage

5, ranges in thickness from c.2 m in the central and eastern part of the Loess Plateau to c.8 m in the west. Whereas in the east it cannot be readily subdivided, in the west, it consists of at least three subdivisions. Moreover, the transitional nature of the climate in the western part of the plateau makes it especially sensitive to the subtle interplay between different depositional, ecological and pedogenic regimes (Li *et al.* 1988) upon which the link between climatic change and its expression in loess stratigraphy depends.

Detailed studies of the magnetic property record from the far western edge of the loess plateau are rather sparse. Here, the loess has a coarser mean particle size as well as a consistently more rapid rate of accumulation, and palaeosols, despite their greater thickness, are often visually less distinct. The present study is based on the major Dadongling section close to Xining and some 270 km west-north-west of Lanzhou (Fig. 1).

The Dadongling Section

The Dadongling section is located in the Xining valley basin and attains a total thickness of 263 m. It is situated on the sixth terrace of the Huangshin, a large tributary of the Yellow River. There are more than twenty palaeosol layers in the section ranging in age from early Pleistocene to Holocene (Li *et al.* in press). The loess and palaeosol sequence formed since the

beginning of the last interglacial is 25 m thick and the last interglacial palaeosol itself (S_1) is 8 m thick.

Palaeosol S_1, (16.9 - 24.8 m), consists of two main palaeosol sub-layers (S_1S_1 and S_1S_2) and a loess layer between them named S_1L_1 (Fig 1). This forms the main loess layer within S_1 throughout the western part of the Loess Plateau, though it cannot be distinguished in sections from the more humid areas in the Central and Eastern parts of the Plateau. It is 1.8m thick and lies between 20.2 m and 22 m. It is yellowish in colour, loose, porous and coarse in texture. These features are typical of the Malan loess of the last glacial. Other features such as the formation of gypsum crystals and the greater abundance of root and earth worm holes are less typical.

Below loess layer S_1L_1 palaeosol S_1S_2 is 3 m thick and lies between 22 m and 25 m in the section. It is the best developed palaeosol in S_1 as a whole and one of the few with signs of strong pedogenesis in the Dadongling section. It is brown to dark brown in colour, blocky in structure and shows signs of strong carbonate leaching. Within this palaeosol, there are also differences in colour, particle size and structure with depth – a more loessic layer ($S_1S_2L_1$) separates an upper light brown, relatively coarse grained palaeosol ($S_1S_2S_1$) from the more strongly developed, lower palaeosol, $S_1S_2S_2$.

Above the loess layer S_1L_1, palaeosol S_1S_1 is 3.3 m thick and lies between 16.9 m and 20.2 m in the section. It, in turn, can

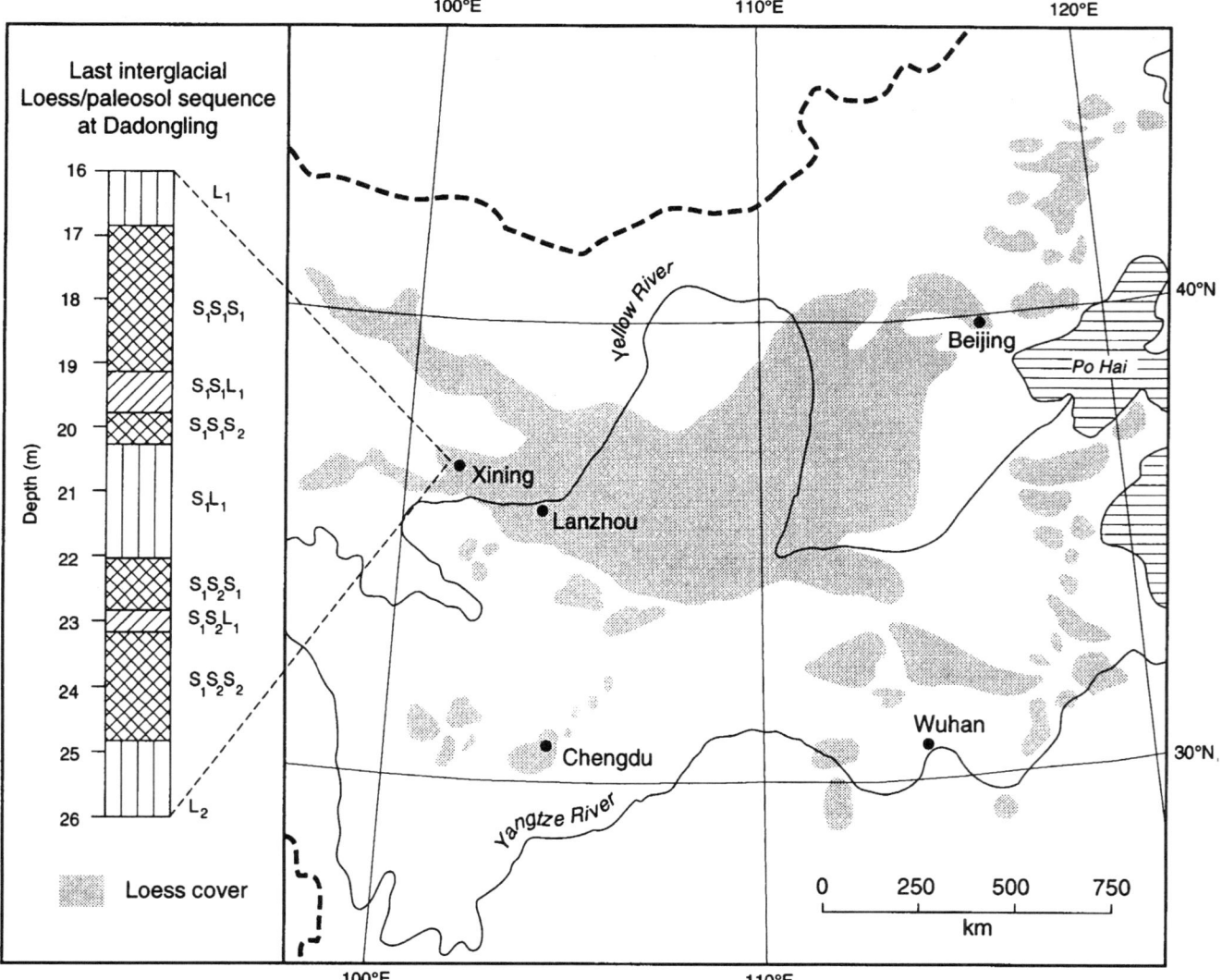

Figure 1. Location of the Dadongling section near Xining. The inset diagram on the left shows the detailed alternation of Loess and Palaeosol units representing the S_1 complex in the section.

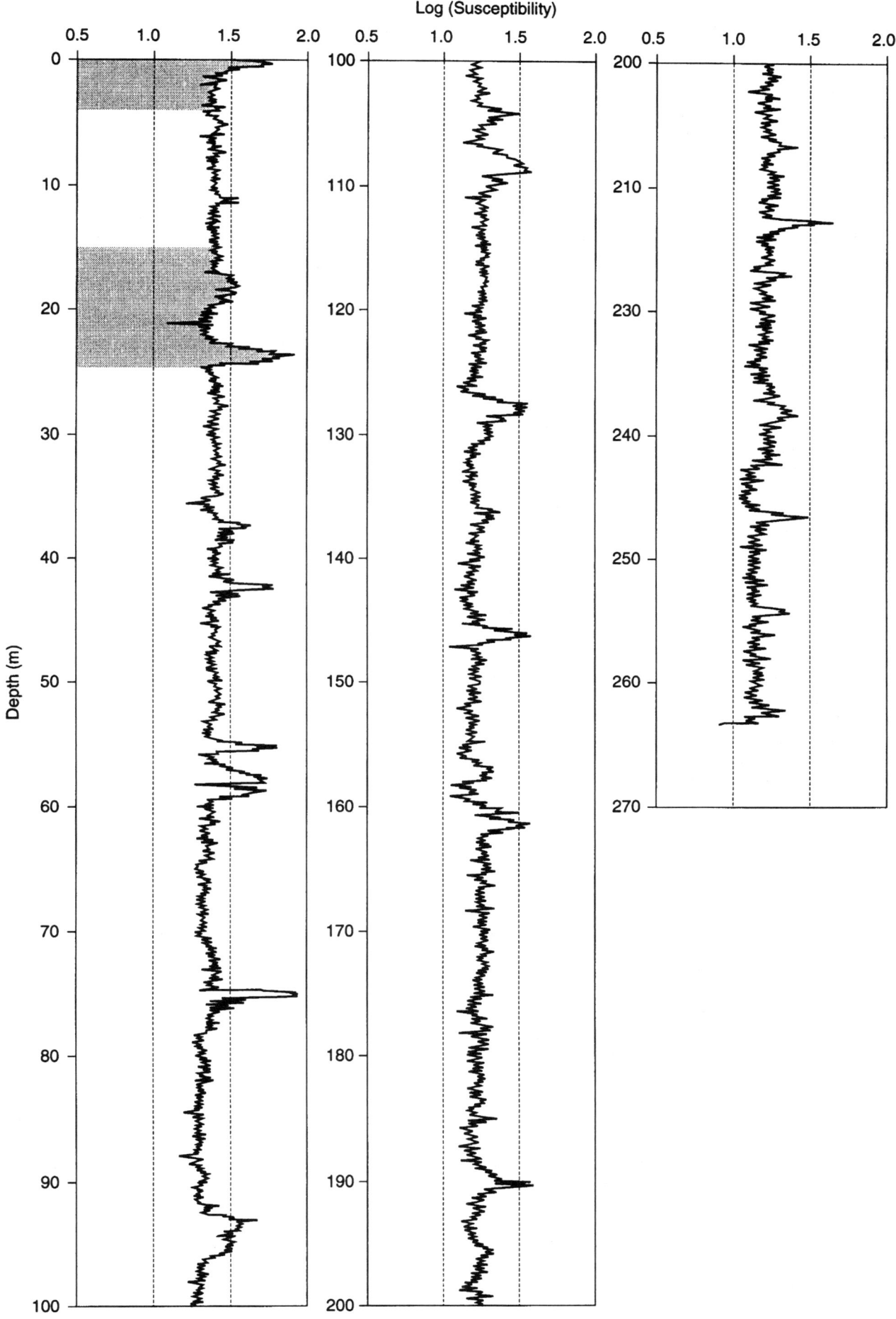

Figure 2. Bartington susceptibility probe readings from the whole of the Dadongling Section plotted on a log scale. The depth intervals considered in the present paper come from the top 25m of the 263m deep section and are shaded.

be subdivided in the field into three further sub-layers, on the basis of colour, structure and grain size. The middle unit of these, classified as a 'loess' ($S_1S_1L_1$), still shows signs of weak pedogenesis in the form of small gypsum crystals, and passes into the palaeosols above and below without any sharp boundary. The uppermost contact of S_1 suggests a rapid end to pedogenesis brought about by high levels of dust deposition.

The S_1 palaeosol sequence as described from the Dadongling Section is representative of all the major sections for the West Loess Plateau and compares closely with that recorded in four sections near Lanzhou some 200 km to the east. The variations recorded in the section many therefore be expected to reflect broad scale regional changes and to bear comparison with other fine resolution records, of continental and global scale palaeoclimate, as confirmed by Li et al. (1992) in their comparison with the Vostock Ice core record in Antarctica.

The L_1 loess, between 16.9 m and 0.77 m is relatively uniform in appearance, light yellow in colour and quite loose. It is much coarser than that in other regions, with the fine sand fraction as much as 35%. Vertical cracks form the main structure. At around 11.5m there is a weakly developed paleosol (not considered in the present study) without clear boundaries above and below. Only its slightly darker colour, finer grain size and higher organic content distinguish it form typical L_1 loess. Similar, but much narrower bands less than 10 cm thick can be distinguished around 1.34-1.39 m and 2.01-2.05 m in the monoliths recovered from the section.

Malan loess in the Xining basin, and especially in the Dadlongling section, is thinner than in other parts of the Western Loess Plateau, for example around Lanzhou. High wind strength in the basin during the last glacial is indicated by the grain size results (Fig. 9). Fine dust once deposited may have been resuspended and transported to other regions . The effect of the strong winds on the section may have been reinforced by its relatively high altitude and exposure within the Xining Valley basin.

The top of the section is modern soil, about 0.2 m to 0.5 m thick. Below it is **Palaeosol S_0**. Its thickness varies from 0.5 m to 1.2 m. It is dark in colour, with a blocky structure and numerous voids. Small $CaCO_3$ nodules can be seen in the field at the base of the soil.

Field Measurements and Sampling

The whole section of 263 m. has been logged at 5 cm. intervals using a field probe linked to a Bartington susceptibility meter. The results are shown in Figure 2. The upper 25 m. of the sections were then sampled by cutting overlapping blocks c.10 cm x 10 cm x 30 cm deep from the face of the outcrop. The blocks (148 in all) were wrapped individually, boxed and transported to the University of Liverpool where they were subsampled for both palaeomagnetic and rock magnetic measurements, as well as for particle size analysis. In most cases the blocks survived intact and could be subsampled in detail, though a minority, especially those from palaeosol horizons, were badly cracked and had partially disintegrated before they could be subsampled. The two depth intervals chosen for detailed rock magnetic characterization are distinguished by shading on Figure 2. The lower one was subsampled by taking contiguous 4-6 cm. thick 'slices'. The upper one was subsampled at intervals varying from 1cm to 5 cm using thinner (1-2 cm) slices. Sample intervals were closest where initial measurements pointed to rapid changes, or where the poor and deteriorating quality of the sample monoliths seriously limited opportunities for critical

subsampling at a later date. In some parts of the upper section, therefore, magnetic measurements have been carried out on contiguous 1 cm slices.

Chronology

The palaeosol S_0 in the West Loess Plateau has been dated in detail by ^{14}C in many sections (Chen et al. 1993). The results show that the palaeosol formed between 3500 BP and 7500 BP, with evidence of an interruption in pedogenesis during the period at some sites. The period of pedogenesis thus broadly corresponds with the Holocene 'climatic optimum'. A ^{14}C age of 5400 ± 50 BP from the middle of palaeosol S_0 at Dadongling is consistent with these previous results.

It is more difficult to date directly palaeosol S_1. The most accepted TL dates for the end of palaeosol formation are around 80 Ka BP (Lou, pers. comm.). Moreover, at the Juizhoutai section near Lanzhou, the T.L. date for the end of S1 is 81Ka BP (Chen et al. 1989). Palaeomagnetic measurements provide an additional chronological marker. The Blake event, one of the main events or excursions in the Brunhes Normal Polarity Epoch, occurred at c.115 ka BP to 120 Ka B.P. during the last interglacial (Tric et al. 1991). Palaeomagnetic measurements from the Dadongling section (Chen et al. in prep.; Zhou, pers. comm.) record the Blake Event close to the base of $S_1S_2S_1$. It is thus possible to correlate this part of the section with oxygen isotope stage 5e. It is suggested that those parts of S_1 between c.23 m and c.24.6 m (subzones $S_1S_2L_1$ and $S_1S_2S_2$) date to between 120 Ka BP and 128 Ka BP and those between c.17 m and c.23 m ($S_1S_2S_1$ to the top of S_1) from 120 Ka BP to 80 Ka BP. These dates suggest mean net accumulation rates for S_1 of between 0.15 mm and 0.2 mm per annum. The provisional chronology proposed for the top 25 m of the section as a whole points to mean rates of accumulation of just under 0.2 mm p.a. This implies that a sample interval of 5 cm provides a temporal resolution of c.250 years, one of 1 cm a temporal resolution of c.50 years.

Laboratory Methods

The samples were all subjected to a sequence of magnetic susceptibility and laboratory remanence measurements. Representative samples were then taken for particle size separation and the same sequence of measurements carried out on the particle size fractions. Finally some 40 percent of the samples from the lower suite analysed were particle sized and the results compared with their magnetic properties (Fig. 7). Details of all the techniques used are set out in the Appendix.

Results

Bulk magnetic properties

Figure 3 summarises the results of the rock magnetic characterization of 337 samples from the two depth intervals indicated on Figure 2. The upper one includes the upper part of L_1 and the undisturbed parts of S_0. The lower one covers the whole of the S_1 sequence and the lower part of L_1.

Close correspondence between the magnetic property variations and the subdivision of the profile into loess and palaeosol units and sub-units makes it possible to use the latter as a basis for describing the rock magnetic stratigraphy of the profile (Fig. 3). In some cases, additional magnetically distinctive horizons have been separately identified with letters. Three main groups of magnetic signature can be identified.

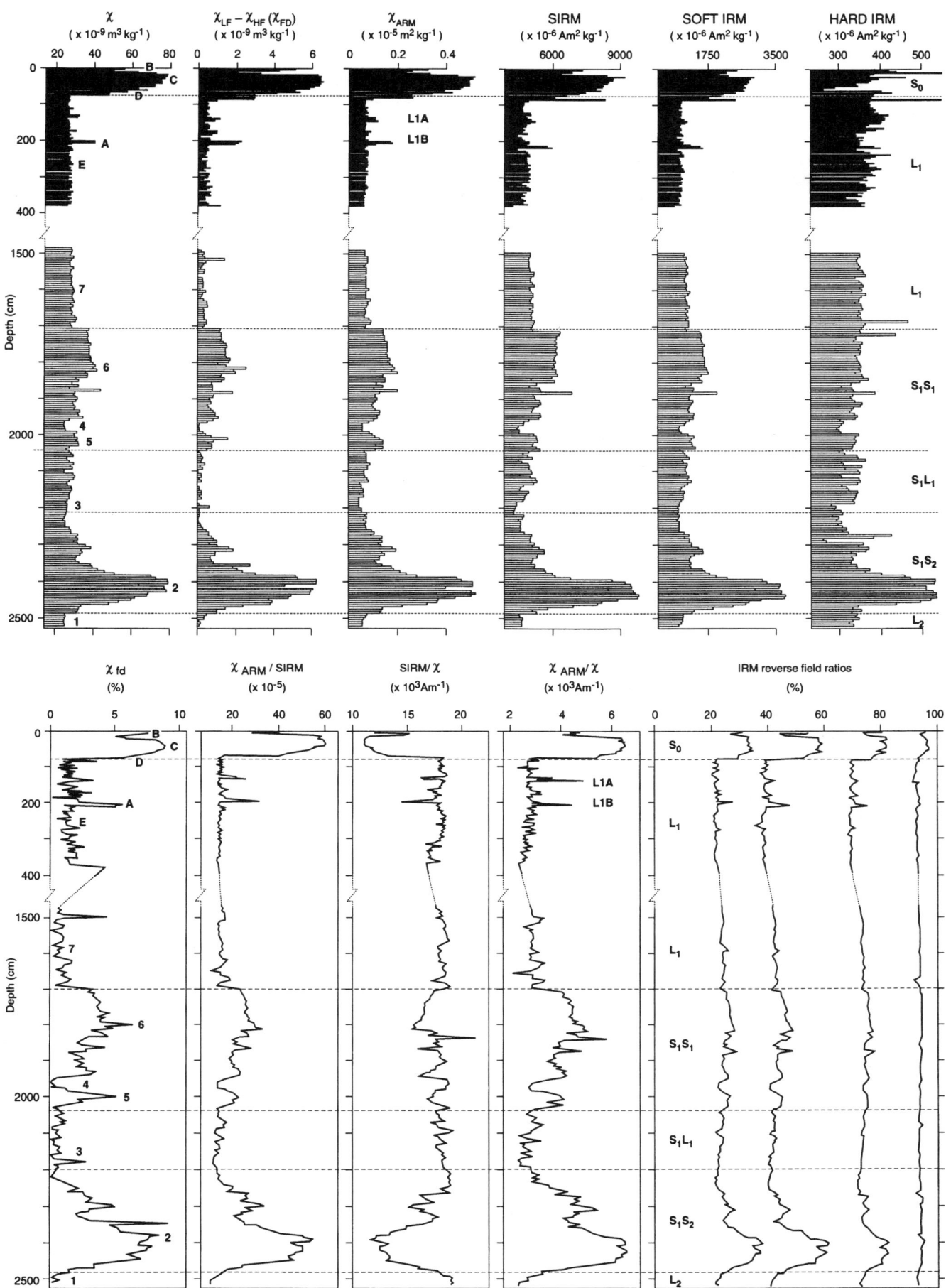

Figure 3. Dadongling Section. Bulk magnetic properties for all the samples measured. The properties used are described in the text, as are the various subdivisions and horizon labels used. The lower part of each plot corresponds with the stratigraphic column shown in the inset of Fig. 1 and also in Fig.7. Depth intervals from which samples have been taken for particle size separation and subsequent magnetic measurements (Figs. 4-6) are also shown. Figure 3a plots mass specific properties, 3b normalised quotients and percentages.

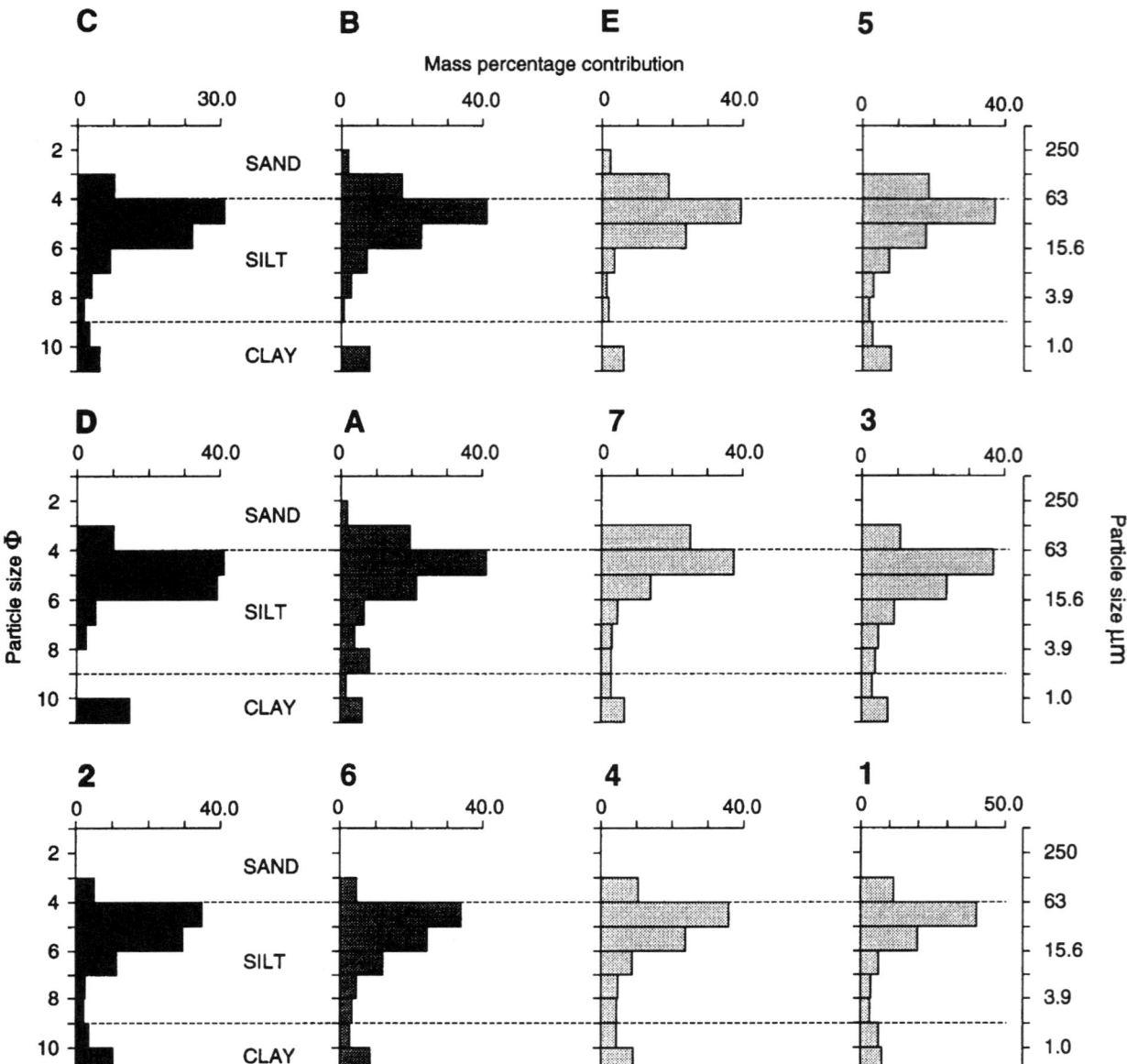

Figure 4. The results of particle size separation by seiving and pipette analysis (see text) on 12 selected samples from the Dadongling section. Samples C, D and 2 come from Palaeosols, (Group 1), samples E, 1,3,4,5 and 7 from typical Loess (Group 2) and samples A,B and 6 from Group 3 samples (see text and Fig 3).

Group 1:- This group includes the samples within palaeosols $S_1S_2S_2$ and S_0. They are distinguished by peak values for χ, χ_{fd}, SIRM, χ_{arm} and 'soft' IRM, as well as for $\chi_{fd}\%$ and $\chi_{arm}/$SIRM. SIRM/χ values are reduced and the reverse field ratios are consistently 'soft'. The hard remanence component peaks consistently in $S_1S_2S_2$ but not in S_0, and this is the only major difference in bulk magnetic properties between these two episodes.

Group 2:- Most of the upper and all of the lower parts of L_1, most of $S_1S_2L_2$, S_1L_1 and $S_1S_1L_1$, and the few basal samples from the top of L_2, all show similar features which are consistently the opposite of those outlined for Group 1. Almost all of L_1 shows quite remarkably uniformity of magnetic properties. The Group 2 samples lie entirely within parts of the section characterised as loess.

Group 3:- The samples from the top of the undisturbed part of the section at the top of L_0, features A & B within the upper part of L_1, most of $S_1S_1S_1$ and parts of $S_1S_1S_2$ and $S_1S_2S_1$, all show characteristics which are variable but always intermediate between those of the other two groups in all respects.

Particle size based measurements

Twelve depth intervals were chosen for particle-size-based measurements as shown on Figure 3. Since some 20g. were required in each case, the samples normally spanned the depth intervals of more than one of the samples used for the initial measurements. Samples 2, C and D come from zones reflecting palaeosols and are classified under '**Group 1**' (major Palaeosols) above. Samples 1, 3, 4,5, 7 and E come from levels classified under '**Group 2** (Loess), and samples 6, A & B come from '**Group 3**' levels.

Figure 4 summarises the results of the particle size separations upon which the magnetic measurements have been made. In all samples, the ϕ 4-5 (31.25 -62.5 μm) fraction makes the largest contribution to the mass, ranging from 34 to 44%. The ϕ 3-4 (62.5-125μm) fraction makes a more variable contribution ranging from 5 to 25%, as does the ϕ 5-6 (16.7 - 31.25μm) fraction, which ranges from 17 to 35%. These three fractions together form 64 to 89% of each sample mass. By contrast, the combined ϕ9 and ϕ10 fractions (<2μm) and the

ϕ10 alone (<1μm) range from only 5 to 13%. The main contrast between loess and palaeosol granulometry is not in the fine fractions but in the balance on either side of the modal coarse silt (ϕ4 - ϕ5) grade, *i.e.* between fine sand, and medium silt. Even in this respect, there is considerable variation within the loess group, (see also Figs. 7 and 8).

These measurements may be compared with those for the loess and palaeosol sample from Luochuan, particle sized using the same methods (Zheng *et al.* 1991). In that case, in the palaeosol, the ϕ10 contribution was 21% and the ϕ3 - ϕ6 contribution 48%, whilst in the loess sample, they were 10% and 65% respectively. Whereas the ϕ3-ϕ4 contribution to the

Dadongling loess reaches as high as 25%, in the Luochuan loess it contributes only 7%.

The magnetic property plots in Figure 5 show, below the particle size distribution for three samples typical of the three groups already defined, the contribution to bulk χ, SIRM, χ_{arm} and χ_{fd} made by each grade. Since all the values relate to a mass of 100g, they can be directly compared not only between particle size fractions but between samples. Sample 7, chosen to represent the '**Group 2**' loess samples presents the simplest case. For every property save χ_{fd} the contributions are entirely dominated by the medium silt to fine sand fractions with a peak in the ϕ4 to ϕ5 component. This is consistent with complete

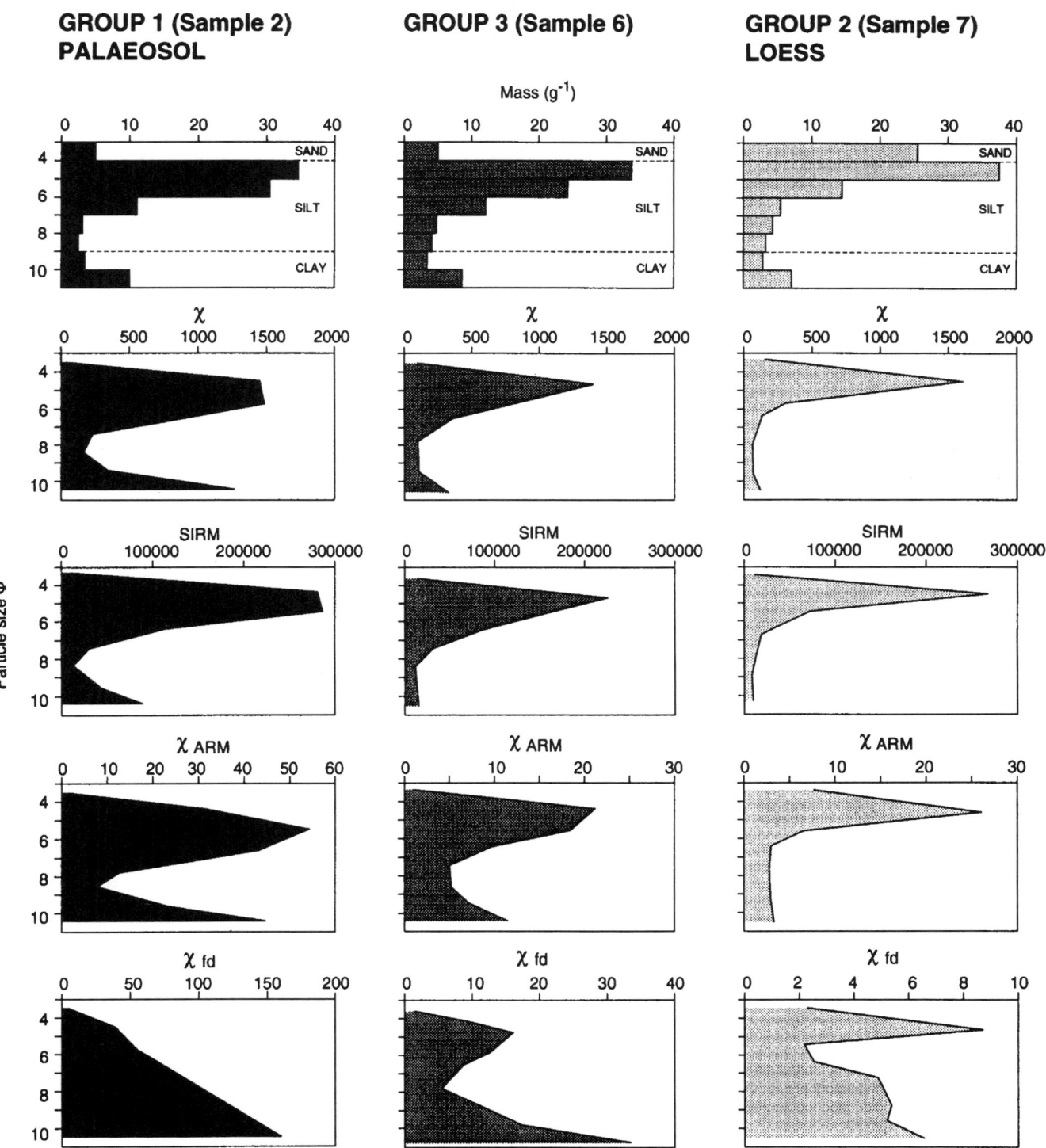

Figure 5. Concentration related magnetic properties for particle size fractions plotted on a 'contribution' basis (see text) for typical Group 2 (2), Group 1 (7) and Group 3(6) samples. Note that the scales for χ_{arm} and χ_{fd} vary from sample to sample.

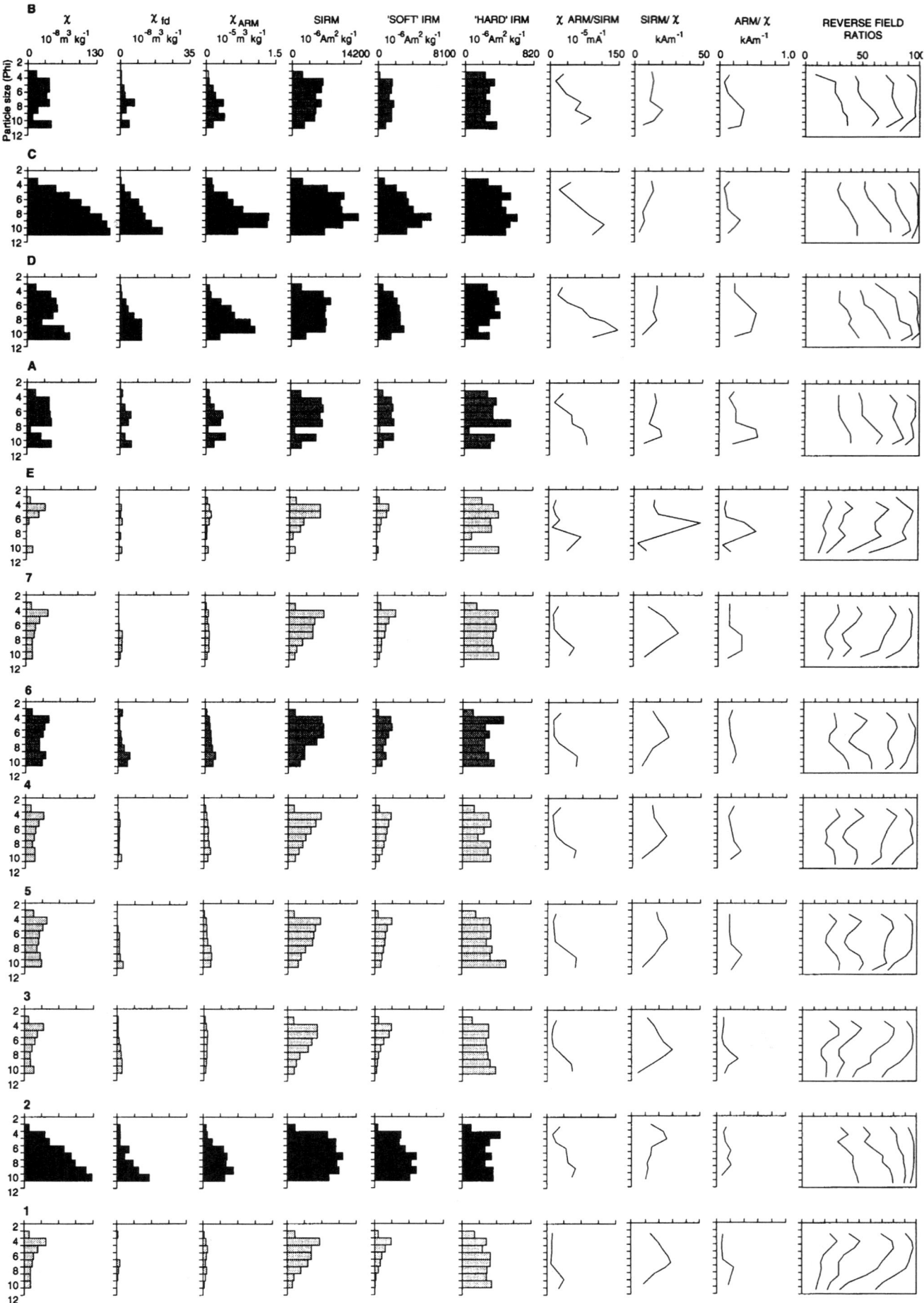

Figure 6. The full range of mass specific magnetic properties and normalised quotients and percentages plotted on a particle size specific basis for the 12 samples shown in Fig. 4 and identified on Fig 3. Shading distinguishes Group 2 (palest) Loess samples from Group 1 (Solid black) Palaeosol, and intermediate Group 3 (Darker stipple) samples. All scales are common to all samples.

dominance of the magnetic properties of these samples by a detrital component associated with aerially deposited medium – coarse silts and fine sands. Similar patterns of particle size contributions are also present in sample 6 typical of '**Group 3**', samples, though the ϕ5-6 contributions to bulk χ_{arm} and χ_{fd} values are close to those for the ϕ 4-5 fraction, and the <ϕ10 makes the main contribution to χ_{fd}. In the '**Group 1**' palaeosol sample 2, the ϕ5-6 contribution exceeds that for ϕ4-5 for all properties and the ϕ10 contributions are more significant especially in the case of χ_{arm} and χ_{fd}. Nevertheless, the detrital ϕ3 to ϕ6 components, taken together, dominate the record even in the palaeosol samples. In this respect, the results contrast strongly with those from the Luo Chuan palaeosol analyzed by Zheng *et al.* (1991).

In Fig 6 the concentration-related magnetic properties for all particle size grades in all samples are plotted on a mass specific basis alongside the quotients and interparametric ratios. This allows consideration of the qualitative variations between particle sizes in each sample, and for each particle size between samples. In the '**Group 1**' samples 2 and C, from the palaeosols, χ and χ_{fd} peak in the finest fractions, χ_{arm} in the ϕ8-10 fractions. χ_{arm}/SIRM values also peak between ϕ9 and ϕ10, reaching values between 60 and 150. Reverse field ratios show an opposite trend to those in many of the '**Group 2**' loess samples, with progressively softer values in finer grades. Sample D and the Group 3 samples are rather similar in most respects and all show strongly bi-modal χ values with peaks in both the finest clay and the main silt and sand 'detrital' sizes. The '**Group 3**' samples all show maximum χ, SIRM and Soft IRM values in the ϕ4-5 range. There is a minimum in χ in the ϕ8-9 size range and often a small peak in χ below this. χ_{arm} and χ_{fd} have low values in all particle sizes and show little consistent trend. χ_{arm}/SIRM values peak below ϕ8 with maxima around 70x10^{-5}. ARM/χ generally peaks between ϕ7 and ϕ9 and SIRM/χ between ϕ7 and ϕ8. Reverse field ratios generally show 'softest' demagnetisation values around ϕ4-6 and hardest high field remanence values below ϕ10.

Rock Magnetic Interpretation

By combining the 'contribution' (Fig. 5) and mass specific (Fig. 6) data, it is possible to propose rock magnetic interpretations for some of the between-grade and between-sample variations. At the outset it is important to recall the limitations of the pipette method of particle size separation as described. All grades between ϕ4 and ϕ10 contain a contribution from finer grades which declines as the grade becomes coarser. Where the mass contribution of a particular grade greatly exceeds that in the clays, the effect is negligible, but where intermediate size factions are poorly represented or only very weakly magnetic, the mass specific values for that grade will be more strongly influenced by 'contamination' from the finest clays. Thus in the case of the present samples, the medium silt to fine sand grades cannot be significantly influenced in this way by clay contributions, but the poorly represented grades between ϕ7 and 9 may be strongly influenced.

Taking the particle sized samples as a whole, there is little evidence for a consistent and significant contribution to magnetic property variations from differences in the hard remanence, presumably haematite or less probably goethite, component. Most between-grade and between-sample variations appear to express ferrimagnetic grain size variations. There is relatively little variation between the loess samples (1,3,4,5,7 and E) and all appear to be dominated in both 'contribution' (Fig. 5) and 'mass specific'(Fig. 6) terms by a

detrital magnetic component in the medium silt to fine sand grades. This mineral phase has a combination of low χ_{arm}/SIRM, χ_{arm}/χ and χ_{fd} values. This points to mean grain sizes coarser than stable single domain (SSD) (Maher 1988), though not uniformly coarse multidomain (MD) in view of the relatively high SIRM/χ values and the intermediate low- backfield ratios.

The palaeosol samples show a similar dominance of the contribution record by magnetic minerals associated with the coarser detrital particles, but there is also an additional phase in the fine grades less 4 μm. This is characterized by peak χ, χ_{fd}, χ_{fd}%, χ_{arm} and χ_{arm}/SIRM values. These are all compatible with ferrimagnetic grains within the SSD and superparamagnetic (SP) range, below 0.10μm. (cf Maher 1988). The strongest indications of SSD grains are in the 1μm - 4μm diameter grades, whereas the finest grades, below 1μm appear to be more predominantly SP. In this respect, values for the Dadongling palaeosols closely parallel those for the Luo Chuan palaeosol (Zheng *et al.* 1991), for which the ratio values all fall within the same range. The presence of significant χ_{fd} values in the 'detrital' silt grades may reflect fine viscous grains on iron coatings.

All the results from the intermediate '**Group 3**' samples are consistent with the interpretation proposed above, in terms of a bi-modal (PSD-MD and SP-SSD respectively) ferrimagnetic grain population. Indeed, the bi-modal trends in χ values in most of the samples reinforce the interpretation, showing peaks matching the χ_{arm} or χ_{fd} peaks in the finest grades, and the SIRM peaks in the coarser.

Stratigraphic Observations

Magnetic properties, particle size variations and the loess/palaeosol stratigraphy

Figure 7 shows, for particle sized samples from the lower of the two sections studied, the relationship between three of the main magnetic grain size indicators, χ_{arm}, χ_{fd}% and χ_{arm}/SIRM, the main trends in particle size variations and the lithostratigraphic log identifying the sequence of loess and palaeosol layers at the site. There is clearly a strong relationship between the magnetic property variation and the alternations of 'loess' and 'palaeosol' layers, though the correspondence is by no means perfect in the case of the narrowest 'palaeosol' layers $S_1S_2S_1$ and $S_1S_1S_2$, where 'peak' magnetic values occupy a significantly narrower depth range than that applied to the palaeosol in the field. There is also a discrepancy at the base of $S_1S_1S_1$ where the increase in magnetic values lies a little way below the defined loess/palaeosol boundary. The special significance of $S_1S_2S_2$, noted from the field stratigraphy, is reinforced by the strength of the magnetic 'signature'. The upper palaeosol layers in S_1 are much less clearly marked save in the changing χ_{fd}% values which appear to provide the most sensitive indication of alternations between periods of recognisable pedogenesis and the intervals between.

Links between particle size variations and magnetic properties are rather less consistent. Pearson 'R' values for the relationship between each grade proportion on the one hand, and eight of the magnetic property values which most clearly reflect the main stratigraphic variations, fail to yield any strong correlations either positive or negative. The only positive correlation greater than 0.5 is between χ_{fd}% and the ϕ7 - ϕ8 proportion (R = 0.5013) and the only negative correlations of comparable strength are between the coarser than ϕ4 fraction and both χ_{fd}% (R = -0.5503) and χ_{arm}/SIRM (R = -0.5213). The highest correlations between magnetic properties and the finer

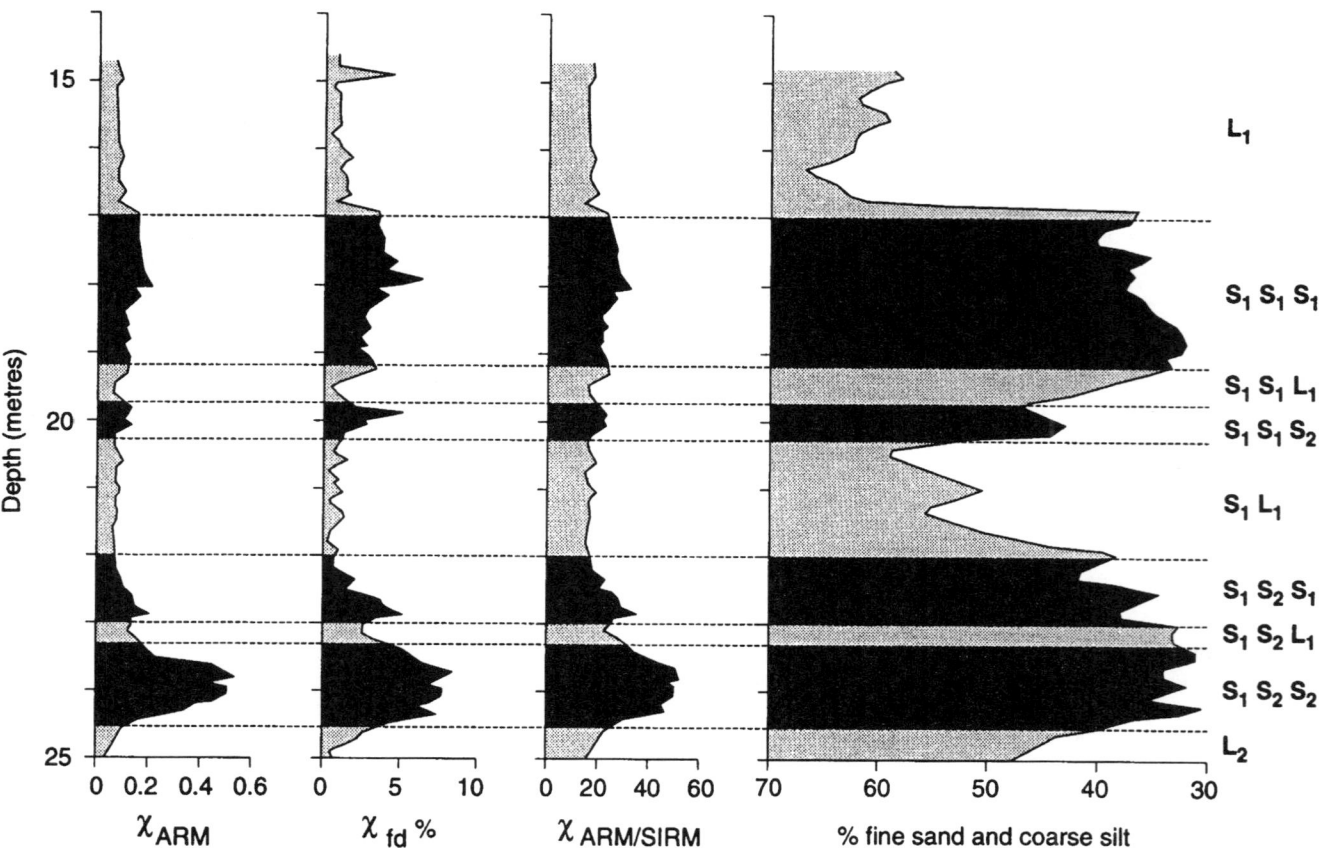

Figure 7. Selected magnetic properties plotted against the main variations in particle size in the lower suite of analyzed samples. The banding relates to the stratigraphic subdivisions shown in Figs. 1 and 3 and discussed in the text.

than $\phi9$ proportion are R = 0.4503 (with χ_{arm}/SIRM) and R = 0.4228 (with χ_{fd}%). A selection of scatter plots (Fig. 8) sheds some light on these observations. In the graphs of χ_{arm}, and χ_{fd}% versus the coarser than $\phi4$ proportion, divergent populations can be seen. Where coarse grade contributions are low and relatively constant, the magnetic property values can vary up to an order of magnitude; where the magnetic property values are low and relatively constant, the fine sand contribution can vary by a factor of 7. This suggests that magnetic property variations are much more sensitive indicators of palaeoenvironmental variations in the palaeosol periods, but are relatively insensitive in the bulk sample measurements during periods of significant granulometric variations *between* the periods of palaeosols formation.

The importance to the magnetic record, of the balance between the contribution from grades coarser or finer than $\phi5$ (31.25μ) in the magnetic record, is reinforced by the total consistency with which every indicator of *fine* magnetic grains, whether concentration related or quotient/percentage based is negatively correlated with variations in contribution coarser than $\phi5$ and positively correlated with variations in contribution finer than $\phi5$. This can be illustrated by the contrasted scatter plots of χ_{arm}/SIRM versus the $\phi4$ - $\phi5$, and $\phi5$ - $\phi6$ fractions respectively. (Fig. 8).

From Figure 7, it is also apparent that changes in the percentage contribution coarser than $\phi5$, the strongest indicator of the main particle size variations, do not respond as consistently to the noted palaeosol/loess transitions as do the magnetic property variations. Whereas the lower contact of $S_1S_1S_2$, the upper contact $S_1S_2S_1$, the lower contact of $S_1S_2S_2$ and the upper contact of S_1S_1S, are clearly marked by major shifts in granulometry, the lithostratigraphic alternations between are not. By contrast, all indicators point to a sudden and very

dramatic shift at the S_1/L_1 boundary. The extreme values for the coarse fraction contribution within 70 cm of the boundary may indicate some resuspension of fines during a period of severe wind stress at the site.

Where the magnetic and granulometric variations are clearly not responding in parallel, as for example during the period of loess deposition within $S_1S_2L_1$, it seems reasonable to infer that the *in situ* climatic thresholds for pedogenesis do not coincide with those controlling variations in the grade of dust delivered from the source regions of the loess. Lack of variation in the latter presumably reflects some constancy in both availability and energy of aeolian transportation, despite the initiation of palaeosol formation at the site. Conversely, during the major periods of loess deposition, lack of variation in the magnetic indicators of pedogenesis, coinciding with significant shifts in grade, may indicate variations in availability or wind energy during periods when local thresholds for pedogenesis were never crossed. These results suggest that by using the magnetic property and particle size records in conjunction, at a range of sites, there is potential for reconstructing the expressions of climatic change with considerable detail and subtlety both spatial and temporal.

The palaeosol horizons near the top of L_1

Within the upper sampled levels of L_1 there are two narrow, clearly defined layers, identified on Fig. 3 as L1A and L1B, that have all the visual and magnetic characteristics typical of brief periods of soil development. One possible explanation for these layers is redeposition of earlier and subsequently eroded palaeosol material from elsewhere. No field evidence has been recorded to support this inference. Moreover, at least one other section from the Western Loess Plateau records a similar

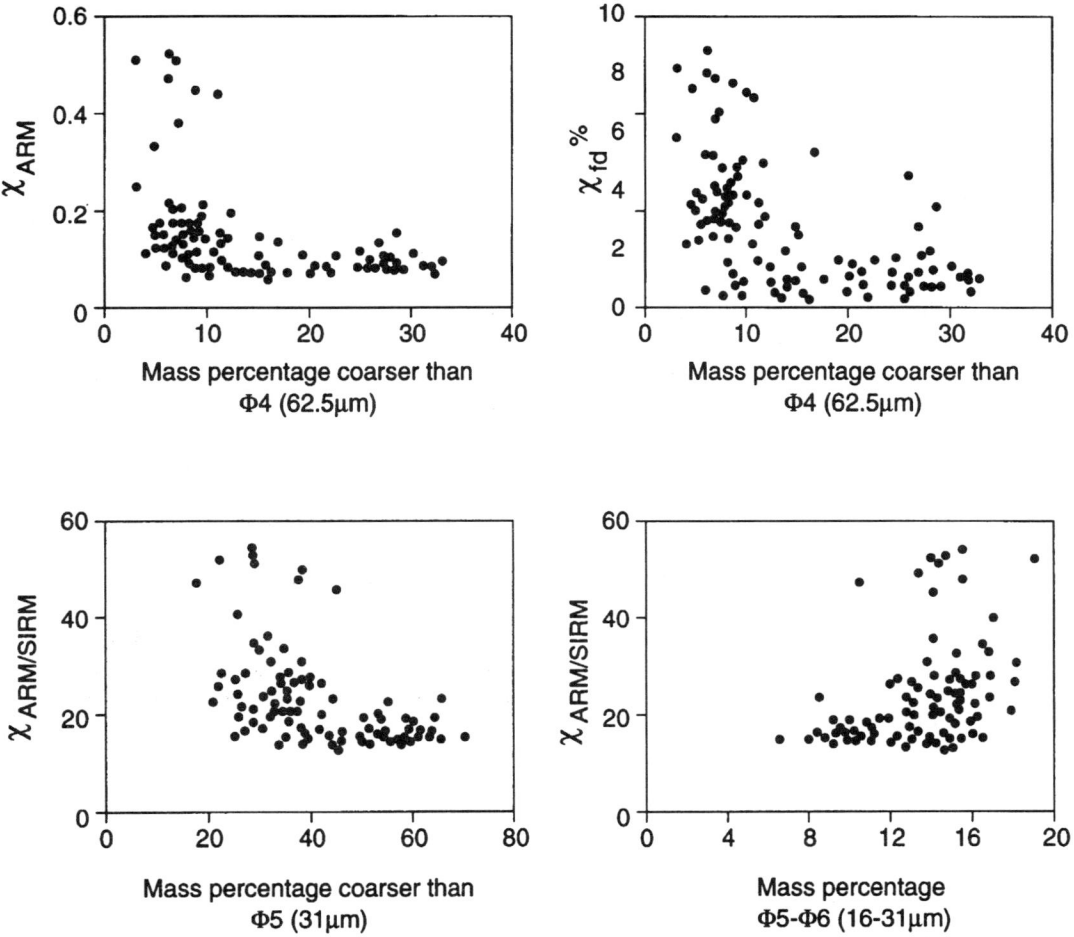

Figure 8. Scatter plots of selected magnetic properties against particle size variables for the lower suite of analyzed samples.

magnetic stratigraphy (Chen, unpublished data). We believe the balance of evidence is in favour of regarding these horizons as representing brief periods of palaeosol development. The upper palaeosol is 5 cm thick, the lower approximately 6 cm thick – it spans the overlap between two monolith blocks, so thickness determination is less precise. Ascribing dates of 80,000 BP and 8500 BP respectively to the base and top of L_1, and assuming a constant rate of accumulation during the period, indicates ages of c.14,300 BP and 11,300 BP for the base of LIB (at 2.05 m) and L1A (at 1.39 m) respectively. These rates of accumulation would indicate a minimum duration of 250-300 years for each period of palaeosol development. Using instead the calculated mean rate of accumulation for S_1 (0.15 mma⁻¹) would indicate that L1B represents c.400 years and L1A c.330 years.

Rates of change and frequency of variation

Some of the changes in magnetic grain size and mineral assemblages point to dramatic shifts over 1-4 cm depth. Accepting, for the present, the inference of syndepositional property variation noted above, and using the mean net accumulation rates already proposed as likely minima in view of the possibility of winnowing in some of the 'glacial episodes,' changes over this narrow depth range imply changes in both directions between episodes of magnetically characterized pedogenesis and non-pedogenesis on timescales of c.50 - 250 years at most.

From the top of $S_1S_2S_2$ to the base of L_1 over a depth interval of just over 6 m, there is evidence for relatively high frequency

variations in magnetic properties indicative of different degrees of supposedly syndepositional pedogenesis. These are most clearly seen in the ARM values and thus ARM/χ and χ_{arm}/SIRM quotients, but other properties independent of these, eg, χ_{fd} (allowing for the difficulty of precise measurements) and the low field (-20 mT and -40 mT) reverse field ratios show largely parallel changes. Using the inferred chronology, the peak to peak period could be as low as 2000 years in parts of the sequence, though the rather coarser approach to sampling in this lower part of the section may have masked finer structure in the signal. At this stage, in view of the relative weakness of the chronological information no attempt has been made to explore questions of frequency.

Despite great uncertainties about dates, timescales and frequencies, the evidence for swift changes and for rather rapidly alternating states, coupled with the previously demonstrated parallelism with the Vostock Ice Core record (Li et al. 1992) holds out the possibility of extracting from this type of section, records approaching the level of temporal resolution achieved in ice core studies.

Conclusions

1. Bulk magnetic measurements from the upper part of the Dadongling section depict a detailed sequence of fluctuating degrees of pedogenesis superimposed on an apparently uninterrupted flux of magnetic minerals associated with the aerially deposited silts and fine sands. These rock-magnetic variations closely parallel the field stratigraphic records, with only minor discrepancies.

2. The main variations in the rock magnetic record reflect variations in the assemblage of ferrimagnetic grain sizes present within each sample.

3. Transitions between rock magnetic 'régimes' often appear to be quite sudden despite the close sample interval and rapid rates of accumulation.

4. There is no clear indication that at this rapidly accumulating site the diagenetic transformations of iron compounds associated with weathering and pedogenesis transgress below the levels contemporary with their operation, and therefore no indication that they 'overprint' the original magnetic signatures below them. The magnetic properties recorded, whether reflecting primary aeolian deposition or secondary formation, may therefore be tentatively regarded as roughly syndepositional.

5. The palaeosol sections representing S_0 & S_1 show broadly similar bulk magnetic properties save that the earlier episode includes evidence for a relatively more important contribution from hard remanence components, presumably haematite or less probably goethite.

6. The dominant particle grades in all the 12 samples studied lie between $16\mu m$ and 125μ and the magnetic properties of these grades largely dominate the magnetic signature of the bulk samples, even in the palaeosols.

7. In the loess samples that show the least indications of any syndepositional pedogenesis, the magnetic properties of the bulk sample reflect a unimodal, relatively coarse PSD-MD ferrimagnetic grain size assemblage.

8. In the palaeosol samples, this coarser ferrimagnetic grain size mode is also represented, but its expression in the bulk sample properties is strongly modified by a fine, (SP-SSD) mode reflecting secondary ferrimagnetic grains found both within the clay fraction and in the coarser particle fractions probably through conversion of paramagnetic and/or imperfect antiferromagnetic iron coatings on the silt and fine sand grains.

9. Comparison between magnetic property and particle size variations suggests that by combining the two lines of evidence, it may be possible to resolve in some detail the interplay between *in situ* processes and those controlling the flux of aeolian material to the site.

10. Consistent variations in magnetic properties towards the top of L_1 may consitute evidence in the section for brief climatic variations reflecting the 'late-glacial' oscillations now widely documented elsewhere.

11. Evidence both for the rapid changes and for high frequency variations in the degree of 'magnetic pedogenesis' point to the eventual possibility, given an adequate chronology, of resolving the record in the section at a level of detail sufficient to permit comparison with ice core records.

Acknowledgements

We are especially grateful for the help given by Bob Jude in many aspects of this research. The contributions of Fahu Chen and Ruijin Wu to the research were funded by the British Council and the Chinese Academy of Sciences. The field work was supported by NSFC fund 49101015.

References

BANERJEE, S.K., HUNT, C.P. & LIU, X-M. (1993). Separation of local signals from the regional palaeomonsoon record of the Chinese Loess Plateau: a Rock-Magnetic approach. *Geophysical Research Letters,* 20, 843-846.

CHEN, F.H., ZHANG, Y.T., CAO, J.X. & ZHANG, W.X. (1989). The comprehensive study on depositional age of Jiuzhoutai loess section, Lanzhou. *ACTA Sedimentologica Sinica,* 7, 120-125.

CHEN, F.H. ZHANG, W.X., *et al.* (1993). The loess stratigraphy and Quaternary glaciations in Gansu and Qinghai region. *Science Press,* Beijing (in press).

CHEN, F.H. Unpublished data.

EVANS, M.E. & HELLER, F. (1994). Magnetic enhancement and palaeoclimate: study of a loess: palaeosol couplet across the loess plateau of China. *Geophysical Journal, Interiors,* 117, 257-264.

FOLK, R.L. (1965). *Petrology of sedimentary rocks,* Hemphills, Austin, Texas, 246 pp.

KING, J., BANERJEE, S., MARVIN, J. & OZDEMIR, O. (1982). A comparison of different magnetic methods for determining the relative grain size of magnetite in natural materials: some results from lake sediments. *Earth Planetary Science Letters,* 59: 404-419.

KUKLA, G., HELLER, F., MING, L.X., CHUN, X.T., SHENG, L.J., SHENG, A.Z. (1988). Pleistocene climates in China dated by magnetic susceptibility. *Geology,* 16, 811-814.

KUKLA, G. & AN, Z.S. (1989). Loess stratigraphy in Central China. *Palaeogeography, Palaeoclimatology and Palaeoecology,* 72, 203-225.

LIU, T.S. *et al.* (1985). *Loess and environment.* Science Press, Beijing.

LI, J.J., ZHU, J.J., KANG, J.C., CHEN, F.H., FANG, X.M., MU, D.F., CAO, J.X., TANG, L.Y., ZHANG, Y.T. & PAN, B.T. (1992). The comparison of Lanzhou loess profile with Vostok ice core in Antarctica over the last glaciation cycle. *Science in China Series B - Chemistry, Life Science & Earth Science,* 35, 476-488.

MAHER, B.A. (1988). Magnetic properties of some synthetic sub-micron magnetites. *Journal of Geophysical Research.* 94: 83-96.

MAHER B.A. & THOMPSON, R. (1991). Mineral magnetic record of the Chinese loess and palaeosols. *Geology,* 19: 3-6.

MAHER B.A. & THOMPSON, R. (1992). Palaeoclimatic significance of the mineral magnetic record of the Chinese loess and palaeosols. *Quaternary Research,* 37, 155-170.

TRIC, E., LAJ, C., VALOT, J.P., TUCHOLKA, P., PATTERNE, M. & GUICHARD, F. (1991). The Blake geomagnetic event: transition geometry, dynamical characteristics and geomagnetic significance. *Earth and Planetary Science Letters,* 102, 1-13.

ZHOU, L.P., OLDFIELD, F., WINTLE, A.G., ROBINSON, S.G., WANG, J.T. (1990). Partly pedogenic origin of magnetic variation in Chinese loess. *Nature,* Vol. 341. No. 6286, pp.737-739.

ZHENG, H., OLDFIELD, F., YU, L., SHAW, J. & AN, Z. (1991). The magnetic properties of particle-sized samples from the Luo-Chuan loess section : evidence for pedogenesis. *Physics of the Earth and Planetary Interiors,* 68, 250-258.

ZHU, R.X., DING, Z.L., ZHOU, L.P. LAJ, C. & MAZAUD, A. (in prep.). The Blake Geomagnetic Polarity Episode recorded in Chinese Loess.

Appendix

Magnetic measurements

The sequence of magnetic measurements carried out on both bulk samples and particle size fractions was as follows:-

(i) low frequency (0.47 kHz) susceptibility ($\chi_{HF} = \chi$);

(ii) high frequency (4.7. kHz) susceptibility (χ_{HF});

(iii) anhysteretic remanent magnetization (ARM);

(iv) saturation isothermal remanent magnetization at 1 T (SIRM)

(v) isothermal remanent magnetization using successively increasing reverse fields of -20 mT, -40 mT, -100 mT and − 300 mT. DC demagnetization rather than acquisition has been used to characterize components of IRM, since this obviates the need for AF demagnetisation after the measurement of ARM.

Susceptibility measurements were carried out using a Bartington Meter and dual frequency sensor. All remanances were measured using a portable Minispin Slow Speed Spinner Fluxgate Magnetometer. Anhysteretic remanences were grown in a modified Molspin alternating field (a.f.) demagnetizer using a peak a.f. field of 100 mT and a steady direct current bias of 0.04 mT. Isothermal remanences were grown in Molspin Pulse Magnetizers.

The full range of measurements has been used to calculate the following mass specific properties:

(i) (low frequency) susceptibility (χ) ($m^3 kg^{-1}$);

(ii) frequency dependent susceptibility ($\chi_{LF} - \chi_{HF} = \chi_{fd}$) ($m^3 kg^{-1}$);

(iii) anhysteretic remanent magnetization (ARM) ($Am^{-2} kg^{-1}$) expressed as susceptibility of ARM (χ_{arm}) ($m^3 kg^{-1}$) (this figure is obtained by dividing ARM by the DC biassing field used (0.04 mT = 31.84 Am^{-1});

(iv) saturation (*i.e.* 1T) isothermal remanent magnetization (SIRM) ($Am^{-2} kg^{-1}$)

(v) 'Soft' IRM = 0.5 (SIRM -IRM_{-20mT}) ($Am^{-2} kg^{-1}$)

(vi) 'Hard' IRM = 0.5 (SIRM + $IRM_{-300\ mT}$) ($Am^{-2} kg^{-1}$).

Among the concentration-related measurements are some which are used here to estimate the changing relative importance of different magnetic components in the samples:

(i) significant frequency dependent susceptibility values (χ_{fd}) denote the presence of fine 'viscous' ferrimagnetic (magnetite or maghaemite) grains close to the stable single domain (SSD)/superparamagnetic (SP) boundary. In the case of magnetite, this refers to grains around 0.02 μm in diameter; (Maher 1988)

(ii) low field ('soft') isothermal remanence as defined above provides a simple basis for approximating the total concentration of remanence carrying ferrimagnets;

(iii) high field ('hard') isothermal remanence is used here as a basis for approximating very crudely the total contribution of remanence carrying haematite or geothite in a sample;

(iv) anhysteretic remanent magnetization (ARM) can be used to give approximate concentrations of stable single-domain ferrimagnetic grains (approximately 0.02-0.04 μm), provided ARM values are high relative to both low field and high field isothermal remanence (Maher 1988).

The main percentages, normalized ratios and quotients which are used here to identify the changing relative proportions of magnetic components in the samples are as follows:-

(i) χ_{fd}%; by expressing χ_{fd} as a percentage of χ_{LF}' variations in the *relative* contribution of fine viscous grains at the SSD/SP border to the total ferrimagnetic assemblage can be estimated.

(ii) ARM/χ (kAm^{-1}) has been proposed by Banerjee *et al.* (1981) and King *et al.* (1982) as a good indicator of magnetite grain size. Maher's results (1988) confirm that this is so for her synthetic samples across the range from stable single-domain (SSD) to multi-domain (MD) grains. In many natural sample suites, its value is reduced by the effects of both superparamagnetic (SP) and paramagnetic contributions.

(iii) SIRM/χ (kAm^{-1}) in natural samples can be influenced by a much wider range of variables. For example, it will be reduced by increased ferrimagnetic versus imperfect antiferromagnetic contributions, by increased mean grain size from stable single-domain upwards and by increased superparamagnetic or paramagnetic contributions to total χ. Comparison between this ratio and the other ratios and quotients often makes it possible to identify the main contributory causes of variation in SIRM/χ. Any reciprocal relationship between this quotient and χ_{fd}% suggests that it is dominated by changes in the relative contribution of fine viscous grains at the SSD/SP border around 0.02 μm diameter.

(iv) χ_{arm}/SIRM (kAm^{-1}) has been shown by Maher (1988) to be strongly related to magnetite grain size across the SSD-MD range. Peak values occur in true SSD grains (approximately 0.02-0.06 μm in diameter) and decline steeply with increasing grain size.

Particle size separation

Particle size fractions for subsequent magnetic measurement were obtained using the following methods:-

The particle size scales used are based on Folk (1965) and the procedures for separation were as follows:-

1. Dried and disaggregated bulk samples were weighed out in a 100cc beaker and dispersed ultrasonically with 25cc calgon (3.3% sodium hexametaphosphate, 0.7% sodium carbonate w/v) and 25cc deionised water.

2. Each dispersed sample was wet-sieved through a 4φ (62.5 μm) brass sieve and so separated into coarse and fine fractions.

3. The sieve contents were washed, transferred to beakers and dried at 40°C. They were then passed through a nest of sieves using a sieve-shaking machine.

4. Finer grades (silt and clay particles) were separated by the pipette method. The material passing through the 4φ sieve was retained, and transferred to a 500 ml settling cylinder. This was filled with deionised water up to 500 ml, shaken end-over-end and placed in a constant temperature (25°C) water bath. Twenty millilitres of the suspension was then pipetted from the cylinder according to settling times based on Stokes' law. The procedures were repeated several times for each of the particle grades in order to improve accuracy. The pipetted samples were centrifuged at 3000 rpm for 15 min, the supernatant discarded, and the residue dried at 40°C. Sample material from each particle size fraction was then weighed, packed and measured.

Particle size analysis

Particle size analysis was carried out on 117 samples from the lower of the two zones of detailed analysis (14.67 -25.00 m) using a Sedigraph 500ET in the University of Leicester. All samples were first disaggregated using 1% calgon then vibrated in an ultrasonic bath for 10 minutes. Organic samples were also treated in 10% Hydrogen Peroxide H_2O_2. Figure 7 compares particle size data indicative of the main variations in granulometry with magnetic property variations for the same samples.

Quaternary Proceedings No. 4, 1995 41-51
© Quaternary Research Association, Cambridge.

Sedimentological Characteristics and Rare Earth Element Fingerprinting of Tibetan Silts and their Relationship with the Sediments of the Western Chinese Loess Plateau.

M.L. Clarke

M.L. Clarke, 1995 Sedimentological characteristics and rare earth element fingerprinting of Tibetan silts and their relationship with the sediments of the western Chinese Loess plateau, In *Wind Blown Sediments in the Quaternary Record* (Edward Derbyshire). Quaternary Proceedings No. 4, John Wiley & Sons Ltd., Chichester, pp. 41- 51.

Abstract

The source of the constituent silt particles of the central Chinese Loess Plateau has long been considered to be the gobi and sand deserts to the north and northwest. However, the western Loess Plateau lies in the foothills of the vast, high level (5000m) Tibetan Plateau and associated mountain chains. Rare Earth Element analysis of silts from the Kunlun Mountains in northern Tibet show strikingly similar signatures to those of loess from Lanzhou, suggesting that the western Loess Plateau deposits have a significant input of sediment derived from Tibet. Aeolian features in the Kunlun Mountains indicate an effective transport system directing fine-grained sediment into the Qaidam Basin where it may deflated by strong frontal winds associated with the Mongolian-Siberian High Pressure System. A sixty metre loess deposit formed within the local mountain environment at Labrang shows further evidence of silt forming processes in northeastern Tibet.

KEYWORDS: rare earth elements, silt, loess, China, Tibet.

M.L. Clarke, Institute of Earth Studies, University of Wales, Aberystwyth, Dyfed SY23 3DB, Wales.

Introduction

The majority of research on Chinese loess has been undertaken within the middle reaches of the Yellow River in the central Loess Plateau at sites such as Luochuan, Baoji and Xifeng, in close proximity to the Ordos and Tengger Deserts. However, the thickest accumulations of loess in China occur near the city of Lanzhou in the westernmost part of the Loess Plateau, surrounded by the mountains of Tibet to the northwest, west and south and the Tengger Desert to the north (Fig. 1). Lanzhou lies in the rainshadow of the Tibetan Plateau, on a steep climatic gradient from the area affected by the Asian Monsoon to the hyper-arid northwest Chinese deserts. This is well illustrated by comparing the mean annual precipitation which varies from 600mm in Xian to 330mm in Lanzhou and 30mm in Golmud (Zhao 1986). The loess deposits at Lanzhou are up to 335 metres thick and rest upon terrace gravels of the Yellow River. The loess-palaeosol sequences are believed to reflect the relative dominance of the Mongolian-Siberian Anticyclone over the southerly monsoon (An Zhisheng et al. 1991).

The constituent silts of the central Chinese Loess Plateau are thought to derive from the gobi (gravel) and shamo (sand) desert regions of northern China and Mongolia (Liu et al. 1985; Wu & Gao 1991). The western region, around Lanzhou, is believed to have an additional input from the tectonically-active mountain ranges of Tibet which lie to the immediate west (Bowler et al. 1987; Zhang et al. 1991). An intense periglacial environment exists in these mountains (Fig. 2),

which also contain active glaciers. These are the two commonly cited pre-requisites for cold climate silt formation (Smalley 1966; Minervin 1974; Konischev 1987). Aeolian transport of silt particles produced in these mountain and/or desert environments is facilitated by the Mongolian-Siberian Anticyclone, which dominates the climate system of northwestern China during winter, and the southerly summer Asian Monsoon, which dominates the climate of southern and central China in the summer, although its effects are currently weak in Lanzhou. Spring dust storms associated with the Mongolian-Siberian high pressure system currently deflate silt from the Taklimakan and Qaidam Basins, reworking silt in their path as they are accelerated by the high mountains through the Hexi Corridor and across the Yellow River valley and deposit loess up to several centimetres thick in the Lanzhou Basin (Wang Jingtai et al., this proceedings). Silt transport also occurs via the Yellow River which rises on the Tibetan Plateau and flows through the Lanzhou Basin. This paper describes a comparison of the particle size and rare earth geochemistry of silts sampled from two areas of Tibet with the loess taken from the western Loess Plateau around Lanzhou.

Sample Sites

The sample sites were located in three regions of the Tibetan Front: the western Loess Plateau around Lanzhou at Jiuzhoutai, Dawan and Sala Shan; Labrang in northwestern Tibet; and several sites across a transect in northern Tibet stretching from

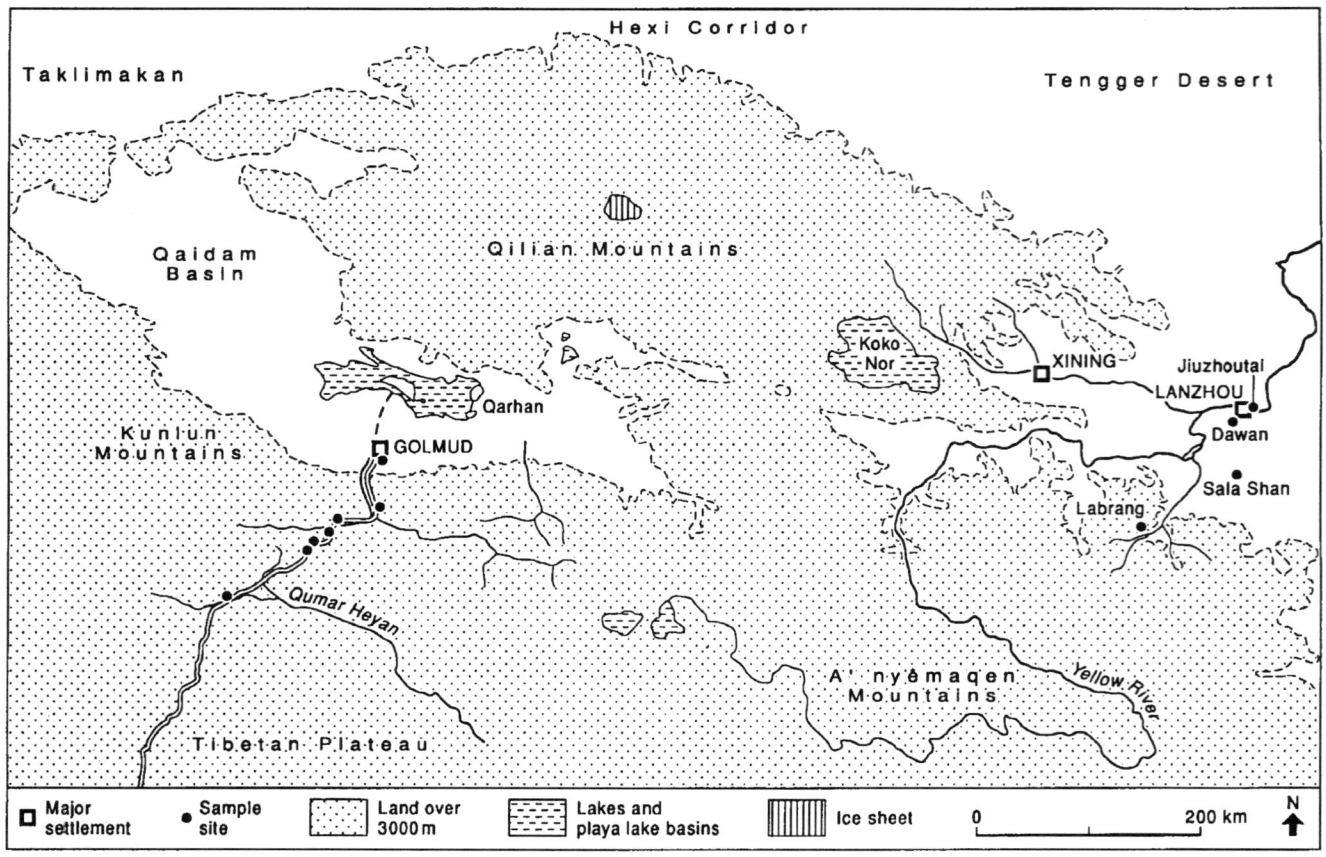

Figure 1. Map of the study area showing locations of samples.

Figure 2. Barchan dune indicating aeolian actitvity in the Kunlun Mountains (4100m), in response to a northerly wind.

the Qaidam Basin, across the Kunlun Mountains and the northern Tibetan Plateau.

Lanzhou

Three sites were investigated around Lanzhou: Jiuzhoutai, Dawan and Sala Shan. Jiuzhoutai lies on the northern bank of the Yellow River in the Lanzhou Basin, overlooking Lanzhou city. It consists of 335 metres of loess resting on terrace gravels of the Yellow River. Loess and palaeosols were sampled from the Scorpion Pit at a depth of 48 to 53 metres (Clarke 1992). In the Scorpion Pit a dual palaeosol is exposed which is thought to be representative of the last interglacial period (Chen *et al.* 1991; Clarke, this proceedings). Dawan section lies on the southwestern edge of Lanzhou Basin, 75km west of Jiuzhoutai, and consists of 273 metres of loess lying on gravels of the Shua Jia River. Primary aeolian loess of unknown age (Clarke 1992; this proceedings) was sampled from depths of 20 to 35 metres along with two triple palaeosol complexes. Loessic alluvium from the Ba River was sampled from the margin of the Loess Plateau at Sala Shan, south of Lanzhou, about half way between Lanzhou and the site in northeastern Tibet at Labrang.

Northeastern Tibet: Labrang

At an altitude of 3000m, within the mountains forming the northeastern boundary of Tibet, there exists a 60 metre loess terrace which extends southeast along the Daxia River Valley from the Tibetan monastery at Labrang (see Clarke 1995, this volume, Fig. 3). The loess has been terraced by the Daxia River and the upper part of this terrace has been cultivated. Sample

blocks were taken from the northern loess terrace about 300 metres east of the Xiahe Binguan. There is no chronological control and no previously published data on this site.

Northern Tibet

The Qaidam Basin contains the highest sand desert in the world, at an elevation of 2600-3000m above sea level. It is directly north of the Tibetan Plateau and surrounded by the Qilian and Altun Mountains to the north and the Kunlun Mountains to the south. It consists of shifting and half-fixed sand dunes resting on gravel gobi formed by aeolian erosion of the Tertiary piedmont plain (Zhao 1986). Yardangs are prevalent in the centre and to the northwest of the Qaidam Basin, indicative of the erosional nature of the environment (Zhu *et al.* 1986). Silt grains, thought to be of Late Pleistocene age, have been found within the halite layers of Qarhan Playa (Chen & Bowler 1987), a salt flat which lies in the central and lowest part of the Qaidam Basin and represents what remains of a much larger Early-Mid Pleistocene freshwater lake (Chen & Bowler 1986). Chen and Bowler (1987) assert that the silt found in the halite layers was generated by glacial and periglacial processes at high altitudes and transported to low level fluvial outwash plains, where it was later deflated by the strong northwesterly winds of the Mongolian-Siberian Anticyclone, and transported southeastward to the Loess Plateau. The area of northern Tibet chosen to be studied here contains several potential silt-forming environments with processes such as cold weathering and glacial grinding dominant in the Kunlun Mountains and on the Tibetan Plateau, and aeolian abrasion (Whalley *et al.* 1982; 1987) and salt weathering (Goudie *et al.*

Figure 3. The Kunlun Pass relict pingo situated an altitude of 4700m.

Figure 4. The surface of the Tibetan Plateau at Qumar Heyan showing the effects of periglacial activity on the Lhasa Highway. Note also the abundance of surface ice patches. The Kunlun Mountains are visible to the north in the background of the figure.

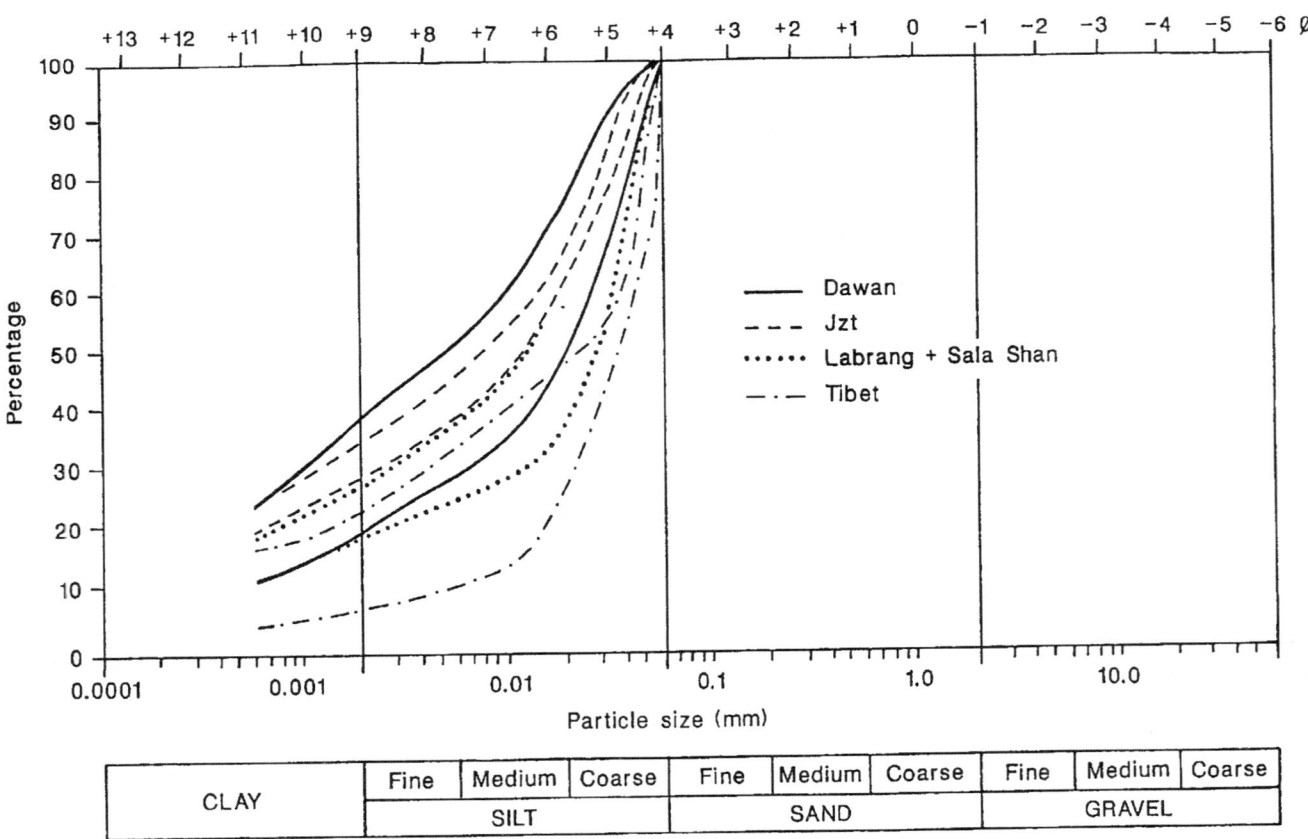

Figure 5. Particle size envelopes for silts from the Tibetan Front.

Table 1. Median diameter of silt samples from sites in the Tibetan Front.
* Samples used for rare earth element analysis.

DAWAN	d^{50} (μm)
Recent soil*	10.0
Loess I	17.8
Loess II*	12.5
Loess III	9.1
Loess IV	14.4
Loess V	11.1
Loess VI	10.4
Palaeosol I	10.8
Palaeosol II	7.7
Palaeosol III	7.8
Palaeosol IV	10.1
Palaeosol V	9.8
Palaeosol VI	5.3

JIUZHOUTAI	d^{50} (μm)
Loess I	8.8
Loess II*	8.3
Loess III	10.2
Palaeosol I	8.4
Palaeosol II	10.0

SALA SHAN	d^{50} (μm)
Loessic alluvium*	17.0

LABRANG	d^{50} (μm)
Labrang B*	29.7
Labrang A	13.3

TIBET	d^{50} (μm)
Kunlun Glacier*	28.7
Kunlun Pingo*	29.8
Qumar Heyan*	26.1
Kunlun Pass*	30.6

1979; Pye & Sperling 1983) in the Qaidam Basin and on the piedmont of the Kunlun Mountains.

The samples taken in this study came from a transect along the Lhasa Highway starting in the Qaidam Basin south of Golmud. Sands were sampled from the Qaidam Basin in order to compare their rare earth element signature with that presented by Wen et al. (1983) for sand from the Tengger Desert. In addition to the Qaidam sand, four samples of silt were taken and one sample of sand from a barchan dune. Buff-coloured (lOYR 7/2) silt was sampled from a surface exposure at an altitude of 3800 metres within the Kunlun Mountains. This sample was labelled 'glacier' as just south of this site was an open area of outwash leading eastwards from the Kunlun River up to a field of unnamed glaciers. The barchan sand sample was taken from one of four northwardly migrating barchan dunes, with crests 8.5 to 10 metres high, at an altitude of 4100 metres to the west of the Lhasa highway (Fig. 2). The remaimng three silt samples comprised of grey silts (2.5GY6/1) from the exposed ice core of a relict pingo just below the Kunlun Pass at 4700 metres (Fig. 3), buff-coloured (10YR7/2) surface silt from the Kunlun Pass at 4767 metres and dull-brown/yellow

(2.5YR7/2) silt from the surface of the Tibetan Plateau at Qumar Heyan (Fig. 4). The surface of the Tibetan Plateau between the Kunlun Pass and Wudoliang is sparsely vegetated and covered with ice patches indicating that moisture is present and suggesting an intensive cold weathering environment. This is also borne out by the freeze-thaw destruction of sections of the Lhasa highway (Fig. 4) ańd the presence of the Kunlun Pass pingo. All of the samples were tested for their rare earth element concentrations and the silts had their particle size distribution measured.

Particle Size

Particle size fractionation across the gobi, the sand deserts and the Loess Plateau, linked with the predominant winds, has been used to account for a desert source for the Chinese loess (Wang 1983). The occurrence of silt on the Tibetan Plateau and in the Kunlun Mountains testifies to the formation of silt in these high mountain environments. Bowler et al. (1987) have argued that silt generated by glacial and periglacial processes at high altitude is transported to low level fluvial outwash plains in the Qaidam Basin, where it is deflated by the predominant northwesterly winds of the Siberian-Mongolian anticyclone and carried to the Loess Plateau. The particle size of the silts sampled in Tibet are here compared to those of loess and palaeosols from the Lanzhou region (described in detail in Clarke 1995).

The samples were measured in a Microtechnics model 5100ET SediGraph which determines the concentration of particles remaining at decreasing sedimentation depths as a function of time. Approximately 3g of sample was disaggregated in a pestle and mortar and dispersed in 0.01% Calgon prior to being pumped into the measurement cell. The cumulative frequency particle size distribution is shown in figure 5 and median particle diameters for the samples shown in table 1.

Rare Earth Element Analysis

The silts from within the mountains to the west and south of Lanzhou were sampled for geochemical fingerprinting (REE) to compare with the work of Wen et al. (1983; 1985) who propose a desert source for Luochuan loess. Geochemical fingerprinting, using Rare Earth Elements, has suggested that the loess deposits within the big bend of the Yellow River were formed within the Tengger Desert (Wen et al. 1983).

Laboratory procedure

Samples from each of the sites were prepared and run on the inductively-coupled plasma atomic emission spectrometer in the Department of Geology at Leicester University. About 0.5g of sample was finely powdered and ignited at 950°C for 1.5 hours prior to digestion in 40% hydrofluoric acid and 67-70% perchloric acid at 180-200°C. When the digestion products attained incipient dryness they were redissolved in warmed 25% hydrochloric acid and made up to 50ml with distilled water before being transferred to the ion-exchange columns. The exchange resin used was a strongly acidic Dowex AG 50W-X8(H), 200-400 mesh, which is a sulphonated gel-type polystyrene resin cross linked with 8% divinylbenzene. After ion exchange, the solutions were then evaporated down to 10-15ml and 2ml of concentrated nitric acid was added. The samples were run on a Philips PV8050 Emission Spectrometer in batches of six including one standard. A pilot project involving silt from the Kunlun Pass and loess from Dawan was

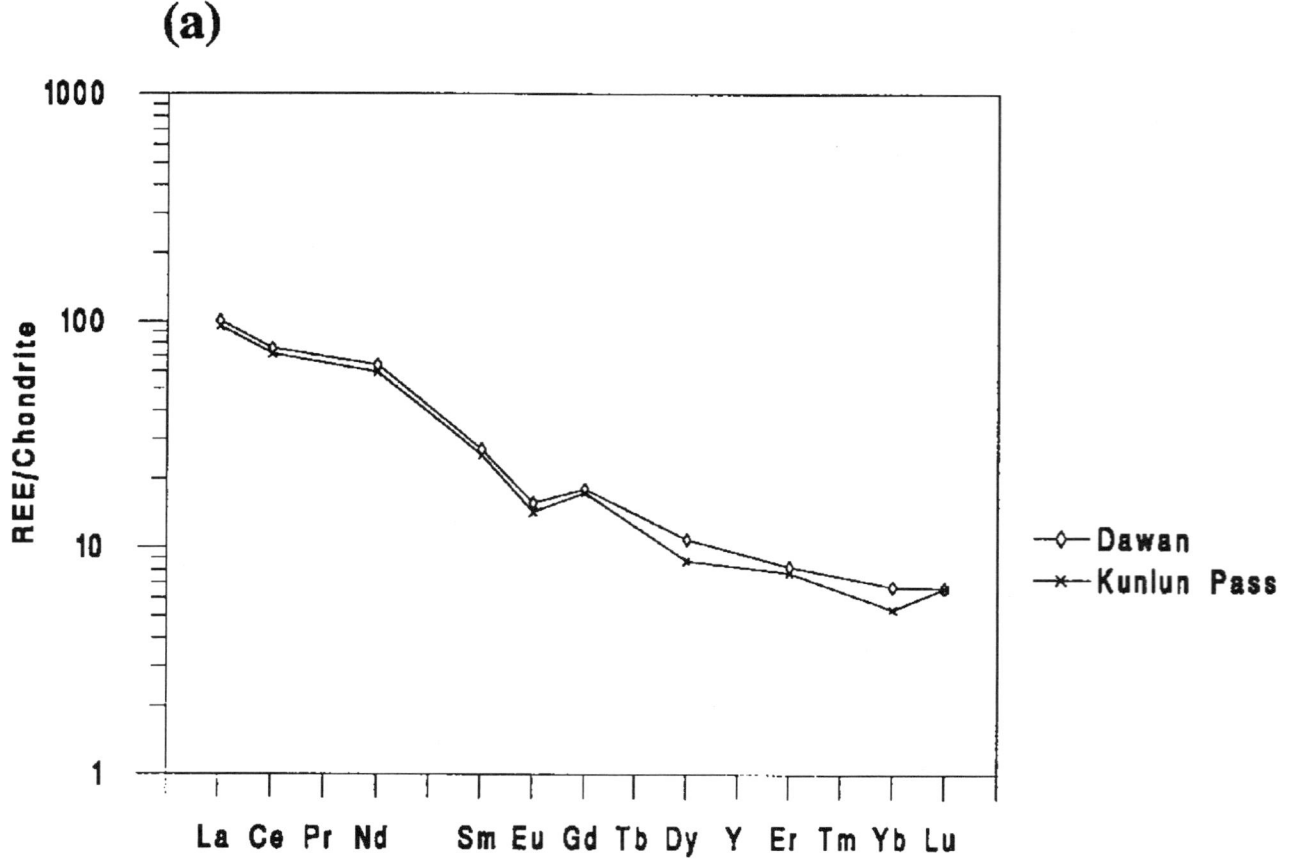

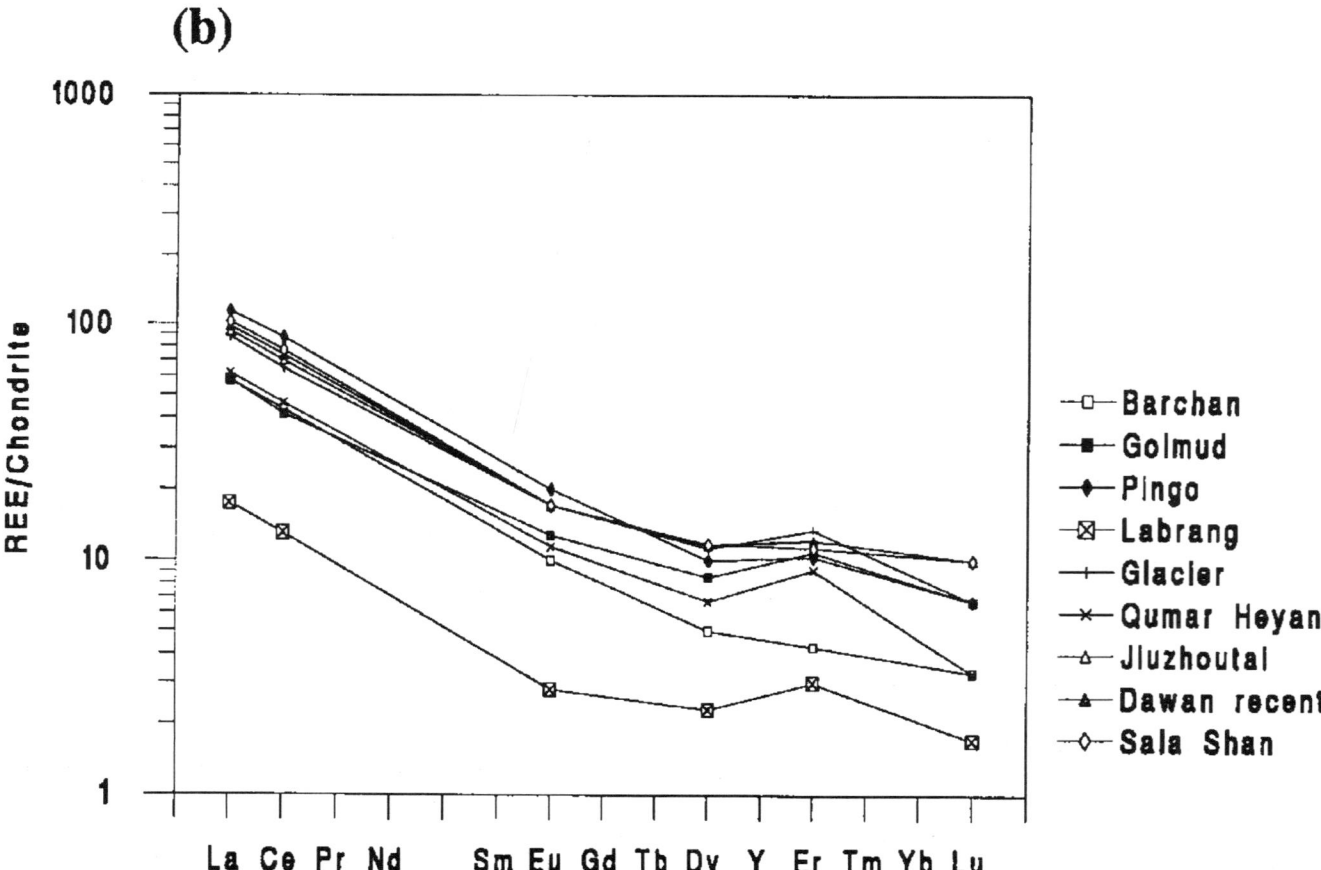

Figure 6a. Chondrite normalised REE distribution for Dawan loess II and silt from the Kunlun Pass; **6b.** Chondrite normalised REE distribution pattern for the other nine sediments from the Tibetan Front.

(a)

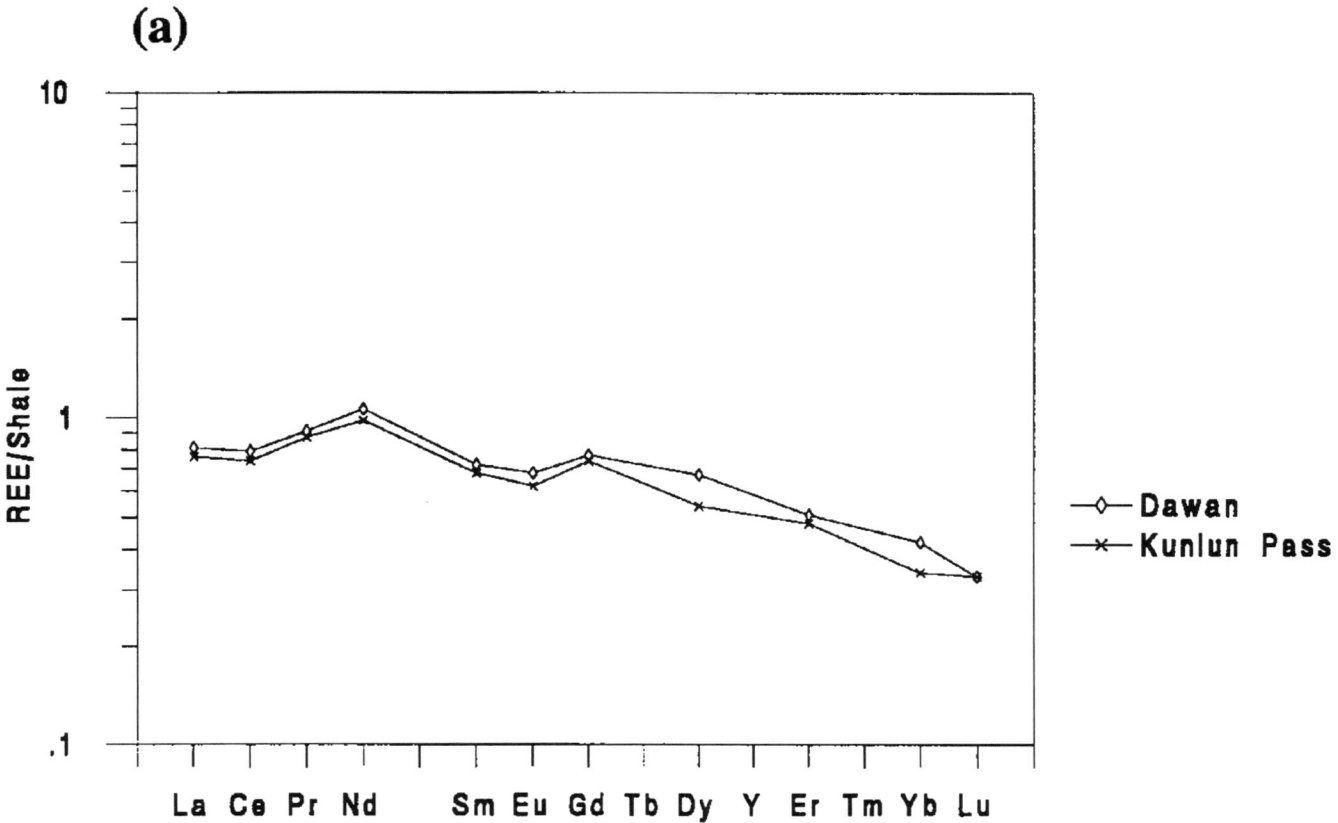

(b)

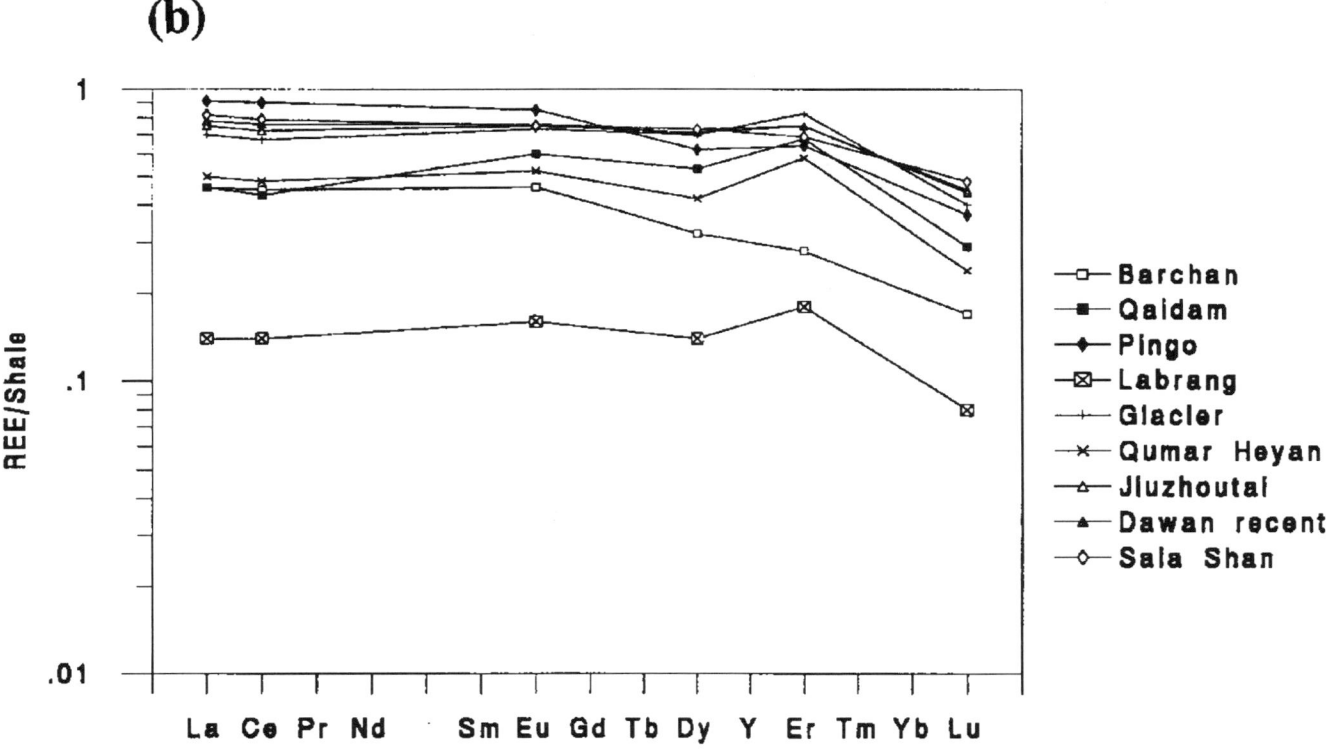

Figure 7a. Shale normlised REE distribution pattern for Dawan loess II and Kunlun Pass silt; **7b.** Shale normalised REE distribution patterns for the other nine sediments from the Tibetan Front.

run one year before the other samples. The similarity of the results from these samples led to a further nine being submitted for analysis, including samples from Labrang, Sala Shan, the Qaidam Basin and recent soil from 1.3 metres below the cultivated surface of a wheat field at Dawan.

Results

The abundance of rare earth elements from the samples is presented in table 2. A check on the performance of the atomic emission spectrometer was made by measuring a laboratory standard of known rare earth concentration; elements which gave anomalous standard readings ($\pm 10\%$) were omitted from the experiment. Thus, the first batch of runs gave anomalous Pr results whereas the second batch of runs involving nine samples showed anomalous Pr, Nd, Gd and Yb results, probably a result of spectral interference. As a result of this difference the two batches of runs are plotted separately.

Chondrite normalised (Nakamura 1974) REE patterns for Tibetan Front sediments are shown in figure 6. They are all enriched in LREE (light rare earth elements: La-Sm) and the Dawan loess and Kunlun Pass silt show negative Eu anomalies. No Ce anomalies were observed in these sediments. The LREE of the nine batch run (Fig. 6b) appear to be arranged into three groups, with silt from the Kunlun Mountains, Sala Shan, Dawan recent soil and Jiuzhoutai loess II forming one group with high LREE abundances. The sand samples from the Qaidam Desert and the Kunlun barchan dune field form another group with silt from the surface of the Tibetan Plateau at Qumar Heyan. The LREE of this 'sand' group is less abundant than that of the 'silt' group by a factor of 1.5-2. The loess from Labrang is particularly interesting in that although the shape of the REE pattern is consistent with that of the other samples, the concentration of REE is 5.5 to 6 times lower. The HREE (heavy rare earth elements: Gd-Lu) distributions follow similar patterns, with the 'silt' samples retaining a higher elemental abundance than the 'sand' (and Qumar Heyan) samples except for the pingo sample which has similar Er and Yb concentrations to the Qaidam Basin sand sampled from near Golmud. The barchan dune sand shows a low Er concentration. The Dawan loess and Kunlun Pass silt samples (Fig. 6a) show REE patterns comparable of the 'silt' group (Fig. 6b).

Figure 7 shows REE patterns for the samples normalised to shales (Haskin & Haskin 1966; see Table 2) which are more indicative of REE abundance from terrestrial sediments. Both Dawan loess II and the Kunlun Pass silt (Fig. 7a) show positively sloped REE patterns from Ce-Nd with Dawan loess II showing a positive Nd peak slightly above the shale value. All the other elements (Fig. 7a) show a decreased abundance with respect to that of the shale. The shale-normalised patterns for the other sediments (Fig. 7b) reflect the three groupings found with the chondrite-normalised patterns. All of the samples tested (Fig. 7a and 7b) showed a lower REE abundance than that of the shale composite. This 'European' shale composite (Haskin & Haskin 1966) was used instead of the North American Shale Composite (Gromet et al. 1984) or the Post-Archean Average Australian Sedimentary Rock (Nance & Taylor 1976) as it was though to be more representative of the crustal average over China.

Discussion

REE abundance is related to mineralogy and particle size, with the >63μm fractions containing less ΣREE than finer fractions with the majority of ΣREE held in particles <20μm in diameter (Wen 1985). The barchan dune sand is probably depleted in REE due to the complete absence of fine material <63μm. The different particle size characteristics of the Qaidam Desert sand and the barchan dune sand probably account for the higher REE abundances in the Qaidam Desert sand, which contains 6.5% of particles <63μm (not shown in Fig. 5). The sample from Qumar Heyan on the Tibetan Plateau (Fig. 5) has a lower REE content than those of the Kunlun and Lanzhou silts despite containing 46.3% of particles less than 20μm, therefore it is possible to argue that the source of this silt may be of a lower REE content rock than that of the Kunlun Mountain silt.

The particle size envelopes in figure 5 show that there is a gradation in particle size from the Tibetan silts to Dawan and Jiuzhoutai loess. This is illustrated by the median values (d50) in table 1. Thus, following the particle size fractionation argument of Wang (1983) it can be argued that there is a link between the sediments of Tibet and the western Loess Plateau. The palaeosols are, in each case, finer than the surrounding loess, with the best developed palaeosols (Dawan III and VI;

Table 2. Abundance of rare earth elements (ppm) in samples of silt and sand from the Tibetan Front. Also shown are the chondrite and shale values used in figures 6 and 7. These are: •an average concentration of REE (ppm) in shales from North America, Europe and the Soviet Union (Haskin and Haskin, 1966); *an average of 10 ordinary chondrites (Nakamura, 1974).

Sample	La	Ce	Nd	Sm	Eu	Gd	Dy	Er	Yb	Lu
Dawan loess	33.3	65.3	40.4	5.4	1.1	4.9	3.7	1.9	1.5	0.2
Kunlun Pass silt	31.4	61.6	37.4	5.1	1.0	4.7	3.0	1.8	1.2	0.2
Qaidam desert sand	19.1	35.8	-	2.9	0.9	-	2.9	2.5	-	0.2
Barchan dune sand	19.0	37.4	-	3.0	0.7	-	1.7	1.0	-	0.1
Glacier silt	29.0	55.6	-	4.7	1.2	-	3.8	3.1	-	0.2
Pingo ice core silt	37.3	75.1	-	5.5	1.4	-	3.4	2.4	-	0.2
Qumar Heyan silt	20.3	39.9	-	2.9	0.8	-	2.3	2.1	-	0.1
Sala Shan alluvium	33.5	66.0	-	5.2	1.2	-	4.0	2.6	-	0.3
Labrang loess	5.8	11.3	-	0.9	0.2	-	0.8	0.7	-	0.05
Jiuzhoutai loess	30.6	59.8	-	4.9	1.2	-	3.9	2.8	-	0.3
Dawan recent soil	32.0	63.0	-	4.9	1.2	-	4.0	2.8	-	0.3
shale•	41.0	83.0	38.0	7.5	1.61	6.34	5.5	3.75	3.53	0.61
chondrite*	0.33	0.86	0.63	0.20	0.07	0.27	0.34	0.23	0.22	0.03

Table 3. Comparison of REE distribution of samples from Dawan and the Kunlun Pass with published results of REE from Luochuan loess (Wen Qizhong *et al.* 1985), Xinji loess (Tian *et al.* 1992), Tengger desert sand (Wen Qizhong *et al.* 1983) and Taklamakan desert sand (Liu *et al.* 1993).

Sample	L a	C e	Pr	N d	S m	E u	G d	D y	E r	Y b	L u
Dawan loess	33.3	65.3	-	40.4	5.4	1.1	4.9	3.7	1.9	1.5	0.2
Kunlun Pass silt	31.4	61.6	-	37.4	5.4	1.0	4.7	3.0	1.8	1.2	0.2
Luochuan loess	31.9	62.4	-	41.3	6.2	1.2	-	-	-	2.7	0.5
Xinji loess	39.0	83.1	-	37.6	6.7	1.3	5.9	-	-	2.7	0.5
Taklimakan sand 1	24.3	43.8	5.14	18.8	3.48	0.72	3.51	2.63	1.46	1.35	0.2
Taklimakan sand 2	8.64	15.1	-	7.59	1.47	0.38	1.18	0.19	0.66	0.65	0.10
Tengger desert sand	3.4	7.7	-	1.2	3.9	-	--	-	-	<1.0	<1.0
Qaidam desert sand	19.1	35.8	-	-	2.9	0.9	-	2.9	2.5	1.2	0.2
Barchan dune sand	19.0	37.4	-	-	3.0	0.7	-	1.7	1.0	0.7	0.1

see Clarke, this proceedings) showing the finest median diameters. Both the loess and palaeosols at Dawan reflect the triplet behaviour mirrored by the magnetic susceptibility of the sequence (Clarke 1995). Mountain loess from Labrang has a coarser grain size than loess at Jiuzhoutai and the majority of Dawan.

The extremely low REE content of the Labrang loess is surprising as the uniformity of composition of loesses from different regions of the world has been previously documented (Taylor *et al.* 1983). It could be due to dilution of sedimentary carbonate minerals which have low REE abundances (Chen *et al.* 1990) although this is unlikely as large quantities of carbonate would be needed to dilute the REE at Sala Shan to the level of Labrang REE. Alternatively, the low ΣREE suggests that Labrang loess derives from a different source than the Loess Plateau silt and is most probably formed within the local mountain environment from a low ΣREE content rock. The similarity of the Sala Shan loessic alluvium to the loess from the Lanzhou sites suggests that it derives from the same source.

Eu/Eu* depletion values (where Eu* = $[(Sm)_N \times (Gd)_N]^{1/2}$ and N stands for chondrite normalised) were available only from the Kunlun Pass (0.68) and Dawan loess (0.71), due to the anomalous Gd signal in the second batch of samples. These results compare favourably with the Eu/Eu* values of 0.63-0.72 derived for loesses from America, Europe, New Zealand and Nanjing (Taylor *et al.* 1983). In wet humid environments Eu^{3+} may be reduced to Eu^{2+} which would be leached out in preference to other rare earth elements increasing the Eu depletion, however the similarity of Eu/Eu* in loess and palaeosols suggests that this did not occur on the loess plateau (Wen *et al.* 1985). Values of $(La/Yb)_N$ (where $(La/Yb)_N$ = [La/Yb)sample /(La/Yb)chondrite]) were also only available for the Kunlun Pass and Dawan samples. These results, Kunlun Pass = 17.5 and Dawan = 14.8, were significantly higher than those of Taylor *et al.* (1983) who found $(La/Yb)_N$ = 7.8 to 11.7 for a variety of loesses. They are also higher than the upper crustal average, documented at 9.3 (Taylor & McLennan 1981). This indicates that the source rocks may have a higher $(La/Yb)_N$ than the crustal average. REE distributions derived from igneous rocks from the Ulugh Muztagh area of northern Tibet (36°28'N, 87°29'E), in the Kunlun Mountains 600km west of the Kunlun Pass, have been found to give high primary $(La/Yb)_N$ ratios of 10.7 to 35.7 (McKenna & Walker 1990). The shape of the chondrite normalised REE distributions obtained from these rocks were similar to the results obtained here for Tibetan Front sediment. Of particular interest were the results from potassium-poor rocks which gave decreased REE abundances in the order

2.8-3.5 times lower than the extrusive and intrusive igneous rocks, although the distribution pattern remained the same. The sample from Labrang showed a REE pattern consistent with that from the other Tibetan Front samples but 5.5-6 times less abundant.

It is possible to compare the REE signature from the Dawan and Kunlun Pass samples with the results of Wen *et al.* (1983; 1985) and Tian *et al.* (1992) for loess from Shaanxi Province which encompasses the central Loess Plateau (table 3). Luochuan is located north of Xian in the big bend of the Yellow River whereas Xinji lies southwest of Xian in the southern margins of the Loess Plateau. The Kunlun Pass and Dawan samples have been chosen as they pose fewer problems with spectral interference than the other Tibetan Front samples. The Xinji data are an average of seven Malan and Lishi loess samples (Tian *et al.* 1992) and the Luochuan data are an average of ten Malan, Lishi and Wucheng loess samples (Wen *et al.* 1985). The LREE of the samples from the Tibetan Front compare favourably with both the Luochuan and Xinji loess with the exception of Ce which is more abundant in the Xinji loess. The distribution of HREE, however, shows a markedly greater abundance in the Luochuan and Xinji loess. It is possible that a variation in source area of these sediments may account for the higher HREE abundances in the central Loess Plateau or that there is a greater abundance of heavy minerals such as zircon, garnet and apatite within the central Loess Plateau as they are enriched in HREE (Taylor & McLennan 1985). The wind drifted sand sample from the Tengger desert (Wen *et al.* 1983) showed a very low REE abundance which is considerably lower than that of the northwestern Chinese and Tibetan sand from the Taklimakan Desert (Liu *et al.* 1993), Qaidam Desert and barchan dune field (Table 3).

Conclusions

Silts from northern Tibet and the western Chinese Loess Plateau show REE patterns steeper than the crustal average and similar to igneous rocks from the Kunlun Mountain range. In addition, the Kunlun Pass silt and Dawan loess show strikingly similar REE signatures, suggesting that the thick loess deposits around Lanzhou have a significant input of sediment derived from the mountain areas of northern Tibet. The presence of a barchan dune field at 4100 metres in the Kunlun Mountains, orientated with crescent arms pointing down valley towards the Qaidam Basin, indicates that there is an effective aeolian transport system directing fine material from the Kunlun Mountains into the Qaidam Basin, where it may be deflated by very strong frontal winds associated with the Mongolian-Siberian high pressure system. The loess from northeastern Tibet at Labrang has a separate source from the Loess Plateau deposits and was probably formed within the local mountain environment.

Acknowledgements

The work presented here was undertaken as part of a NERC studentship (GT4/88/GS/61) and would not have been possible without the enthusiasm and practical assistance of the late Prof. Wang Jingtai of the Geological Hazards Research Institute, Gansu Academy of Sciences. I would like to thank Dr. J.A. King, Dr. C.H. Scott, Meng Xingming and Ma Jinhui for field assistance, and Prof. E. Derbyshire, Dr. J. Shaw, Prof. An Zhisheng, Dr. Zhou Liping and the Lanzhou Institute of Glaciology and Cryopedology for support and encouragement. The REE measurement was carried out in the Geology Department, Leicester University with the assistance of A. Holmes, Dr. B. Dickie and Dr. A. Saunders. This is publication number 378 of the Institute of Earth Studies, University of Wales, Aberystwyth.

References

AN ZHISHENG, WU XIHAO, WANG PINXIAN, WANG SHUMING, DONG GUANGRONG, SUN XIANGJUN, ZHANG DE'ER LU YANCHOU, ZHENG SHAOHUA & ZHAO SHINGLIN (1991). Changes in the monsoon and associated environmental changes in China since the last interglacial. In: Liu Tungsheng (ed) Loess, Environmental and Global Change, Science Press, Beijing. 1-29.

BOWLER, J.M., CHEN KEZAO & YUAN BAOYIN (1987). Systematic variation in loess source areas: evidence from Qaidam and Qinghai Basins, western China. In: Liu Tunghseng (ed) Aspects of Loess Kesearch. China Ocean Press, Beijing, 39-51.

BOWLER, J.M., HUANG QI, CHEN KEZAO, HEAD, J.M. & YUAN BAOYIN (1986). Radiocarbon dating of playa-lake hydrological changes: examples from northwestern China and central Australia. Palaeogeography, Palaeoclimatology, Palaeoecology, 54, 241-260.

CHEN FAHU, LI JIJUN & ZHANG WEIXIN (1991). Loess stratigraphy of the Lanzhou profile and it's comparison with deep-sea sediment and ice core records. Geojournal, 24, 201-209.

CHEN HONGCHEN, JAHN BORMING, LEE T., CHEN CHAOHSIA & CORNICHET, J. (1990). Sm-Nd isotopic geochemistry of sediments from Taiwan and implications for the tectonic evolution of southeast China. Chemical Geology, 88, 317-332.

CHEN KEZAO & BOWLER, J.M. (1986). Late Pleistocene evolution of salt lakes in the Qaidam Basin, Qinghai Province, China. Palaeogeography, Palaeoclimatology, Palaeoecology, 54, 87-104.

CLARKE, M.L. (1992). Formation, depositional history and magnetic properties of loessic silts from the Tibetan Front, China. Unpublished PhD thesis, Leicester University, 164pp.

CLARKE, M.L. (1995). A comparison of magnetic fabrics from loessic silts across the Tibetan Front, western China. Quaternary Proceedings, this volume.

GOUDIE, A.S., COOKE, R.U. & DOORNKAMP, J.C. (1979). The formation of silt from salt dune sand by salt weathering processes in deserts. Journal of Arid Environments, 2, 105-112.

GROMET, L.P., DYMEK, R.F., HASKIN, L.A. & KOROTEV, P.L. (1984). The "North American Shale Composite": it's compilation, major and trace element characteristics. Geochimica et Cosmochimica Acta, 48, 2469-2482.

HASKIN, M.A. & HASKIN, L.A. (1966). Rare earths in European shales: a redetermination. Science, 154, 507-509.

KONISCHEV, V.N. (1987). Origin of loess-like silt in Northern Jakutia, USSR. Geojournal, 15, 331-346.

LIU CONGQIANG, MASUDA, A, OKADA, A., YABUKI, S., ZHANG JING & FAN ZILI (1993). A geochemical study of loess and desert sand in northern China: implications for continental crust weathering and composition. Chemical Geology, 106, 359-374.

McKENNA, L.W. & WALKER, J.D. (1990). Geochemistry of crustally-derived leucocratic igneous rocks from the Ulugh Muztagh area, northern Tibet and their implications for the formation of the Tibetan Plateau. Journal of Geophysical Research, 95, 21483-21502.

MINERVIN, A.V. (1974). Cryogenic processes in loess formation in Central Asia. In: Velichko, A.A. (ed) Late Quaternary Environments of the Soviet Union. Longman, London. 133-140.

NAKAMURA, N. (1974). Determination of REE, Ba, Fe, Mg, Na and K in carbonaceous and ordinaly chondrites. Geochimica et Cosmochimica Acta, 38, 757-775.

NANCE, W.B. & TAYLOR, S.R. (1976). Rare eath element patterns and crustal evolution - I. Australian post-Archaen sedimentary rocks. Geochimica et Cosmochimica Acta, 40, 1539-1551.

PYE, K. & SPERLING, C.H.B. (1983). Experimental investigation of silt formation by dry static breakage processes: the effects of temperature, moisture and salt on quartz dune sand and granitic regolith. Sedimentology, 30, 49-62.

SMALLEY, I.J. (1966). The properties of glacial loess and formation of loess deposits. Journal of Sedimentary Petrology, 36, 669-676.

TAYLOR, S.R. & McLENNAN, S. (1985). The Continental Crust: It's Composition and Evolution. Blackwell Scientific Publications, Oxford.

TAYLOR, S.R., McLENNAN, S.M. & McCULLOCH, M.T. (1983). Geochemistry of loess, continental crust composition and crustal modal ages. Geochimica et Cosmochimica Acta, 47, 1897-1905.

TIAN JUNLIANG, CHOU CHENLIN & EHMANN, D. (1992).

Determination of major and minor trace elements in loess, palaeosol and precipitation layers in a Pleistocene loess section, China, by INAA. *Journal of Radioanalytical and Nuclear Chemistry, Articles*, 110, 261-274.

WANG JINGTAI (1995). Modern sand storms and loess deposition in China. *Quaternary Proceedings*, this volume.

WANG YONGYANG (1983). *Rock desert, gravel desert, sand desert, loess*. Science Press, Beijing, 234p

WEN QIZHONG, YU SUHUA, GU XIONGFU & LEI JIANQUAN (1983). 'Preliminary investigation of REE in loess'. Geochemistry, 2, 81-88.

WEN QIZHONG, YU SUHUA, SUN FUQING, WANG YUQI, CHEN BINGRU, TU SHUDA & SUN JINGXIN (1985). 'Rare-earth elements in Luochuan loess, Shaanxi Province'. *Geochemistry*, 4, 172-180.

WHALLEY, W.B., MARSHALL, J.R. & SMITH, B.J. (1982). Origin of desert loess from some experimental observations. *Nature*, 300, 433-435.

WHALLEY, W.B., SMITH, B.J., MCALISTER, J.J & EDWARDS, A. (1987). Aeolian abrasion of quartz particles and the production of silt-sized fragments, preliminary results and some possible implications for loess and silcrete formation. *In:* Reid, I. and Frostick, L. (eds) *Desert Sediments, Ancient and Modern*. Blackwell, Oxford. 129-138.

WU ZIRONG & GAO FUQING (1985). The formation of loess in China. *In:* Liu Tungsheng (ed) *Quaternary Geology and Environment of China*. China Ocean Press, Beijing, 137-138.

ZHANG LINYUAN, DAI XUERONG & SHI ZHENGTAO (1991). The source of loess material and the formation of the Loess Plateau in China. *Catena Supplement*, 20, 1-14.

ZHAO SONGQIAO (1986). *Physical Geography of China*. Science Press, Beijing.

ZHU ZHENDA, LIU SHU, WU ZHEN & DI XINMIN (1986). *Deserts in China*. Institute of Desert Research, Academia Sinica, Lanzhou.

Quaternary Proceedings No. 4, 1995 53-58
© Quaternary Research Association, Cambridge.

The Citrate-Bicarbonate-Dithionite (CBD) Removable Magnetic Component of Chinese Loess.

Xiuming Liu, Tim Rolph and Jan Bloemendal

Xiuming Liu, Tim Rolph & Jan Bloemendal, 1995 The citrate-bicarbonate-dithionite (CBD) removable magnetic component of Chinese loess, In *Wind Blown Sediments in the Quaternary Record* (Edward Derbyshire). Quaternary Proceedings No. 4, John Wiley & Sons Ltd., Chichester, pp. 53- 58.

Abstract

By analysing and comparing the thermomagnetic characteristics of loess and palaeosols, and synthetic magnetite and maghaemite, both before and after CBD treatment, we conclude that the CBD method is a complex iron-oxide removal process. It is highly unlikely that it can be used to separate the detrital and pedogenic components in loess and palaeosols because not only does the technique apparently dissolve all the iron-oxides but the efficiency of the process is related to the duration of the treatment. A previous suggestion that the CBD method preferentially removes the iron-oxide maghaemite, is not supported by this study. Instead we believe that a single application of the CBD method preferentially removes the ultrafine iron-oxides below an as yet undetermined diameter, but will also dissolve a rim from the outside of the larger oxide grains. The question of what this magnetic fraction represents in loess, pedogenic or primary material, is still open to debate, and the answer will probably include a strong site-dependence related not only to local climatic variability but also to the proximity of the source areas from which the loess is derived.

KEYWORDS: loess, palaeosols, thermomagnetism.

Xiuming Liu and Jan Bloemendal, Department of Geography, University of Liverpool, Liverpool L69 3BX, UK.

Tim Rolph, Geomagnetism Laboratory, University of Liverpool, Liverpool L69 3BX, UK.

Introduction

The magnetic susceptibility record of the Chinese loess-palaeosol sequence is widely regarded as a palaeoclimatic indicator and as such the origin of the susceptibility signal, and its modulation, have been widely investigated. It has been reported from mineral magnetic studies that magnetite, maghaemite (and intermediate compositions; cation-deficient (CD) magnetites) and haematite are the most important magnetic minerals in Chinese loess (Evans & Heller 1994; Heller *et al.* 1991; Liu *et al.* 1990; 1992; Rolph *et al.* 1993). When considering magnetic susceptibility, haematite (α-Fe_2O_3) may be ignored due to its small contribution in comparison to magnetite (Fe_3O_4) and maghaemite (γ-Fe_2O_3). Maghaemite, due to its similar magnetic properties, is usually grouped with magnetite to discuss 'ferrimagnetic' susceptibility (Heller *et al.* 1991; Liu *et al.* 1992). However, a recent paper (Verosub *et al.* 1993) has reported an attempt to separate these two ferrimagnetic components in loess and palaeosols using the citrate-bicarbonate-dithionite (CBD) technique (Mehra & Jackson 1960). The supposition that the CBD method could provide a means of separating magnetite and maghaemite was based on an earlier study by Fine and Singer (1989) of two Californian soil chronosequences, in which the technique proved to be successful at completely removing authigenic soil

maghaemite while leaving magnetite untouched. It is on this premise that the results of Verosub *et al.* (1993) were interpreted, although they concede that ultrafine magnetite, which is commonly considered to be pedogenic in origin, would also be removed by the CBD method. Their results from the Luochuan loess section led them to conclude that the magnetic susceptibility signal of both loess and palaeosols is primarily pedogenic in origin and, furthermore, to believe that the magnetic mineral dominantly responsible for the pedogenic enhancement is maghaemite. If correct, these results require a comprehensive review of current opinion regarding the origin of the climatic signal in loess. It is therefore important to quantify the effect of the CBD technique.

The approach we have used is to apply standard rock-magnetic techniques to both synthetic ferrimagnetic oxides and loess/palaeosol samples, both before and after CBD treatment. These experiments will enable us to judge whether the CBD technique can provide a general method for quantifying pedogenesis in loess.

Samples and the CBD treatment

The CBD method, as a general technique to remove iron oxide from soils and clay, is widely used in the soil sciences. However, this method is not considered to be a special

Table 1. Magnetic susceptibility measurement before and after CBD

	before CBD		after CBD		
sample	grain size (μm)	Susc. (10^{-6} m³kg⁻¹)	Fe** (ppm)	Mass (g)	Susc. (10^{-6} m³kg⁻¹)
M10	SD	3647	106110	1.0213	2954.9
M9	90-120	4378	2174	1.0577	3059.8
M8	90-120.H	2431	2786	1.2115	2010
M2	250-355.H	2213	2093	1.0585	2088.3
M3	355-500.H	2457	2961	1.0649	2320.9
M4	355-500.H		2228		
M5	355-500.H		2745		
M6	355-500.H		2502		
M7	355-500.H		2646		
M4(2)			2970	1.1963	1967.9
M5(2)			2583		
M6(2)			2628		
M7(2)			2273		
M5(3)			2268		
M6(3)			2498		
M7(3)			2241	1.1707	1961.6
M5(4)			1958		
M6(4)			1751	1.2717	1785 .2
M5(5)			2709	1.2903	1797.1

H: heated to 700°C. (2): number of CBD applications
**: AAS measurements from CBD solution.

technique for maghaemite removal and, in fact, was originally aimed at haematite and goethite removal (Mehra & Jackson 1960). Nevertheless, a recent study (Fine & Singer 1988) has indicated that the CBD method can be used to completely dissolve maghaemite whilst leaving all but the finest magnetite grains untouched. This result is contrary to an earlier study where it was reported that soil maghaemite was only slightly soluble with this method (Taylor & Schwertmann 1974). In order to verify whether magnetite is fully resistant to the CBD method, the technique was applied to one gram each of synthetic magnetite powder of four different grain size ranges (see Table 1; samples provided by Dr. W. Owens, Birmingham University). In addition, magnetite with a grain size range between 355-500 μm was used to investigate the effect of repeated CBD treatments on the same sample, and in table 1 the number of treatments is indicated following the sample code. Some samples were also heated to 700°C in order to remove any maghaemite or cation-deficient (CD) magnetite prior to the CBD treatment, and these samples are indicated in table 1 with the letter H.

In applying the CBD method we followed exactly the technique as described by Mehra and Jackson (1960). After each CBD treatment, the solution remaining was collected separately and the sample was put in a dryer at 38°C for 30 hours. After drying, the samples showed a small increase in mass due to the presence of a sticky residue (yellow-green in colour for all samples), and the susceptibility of the all magnetite samples showed a marked decrease. The material dissolved by the treatment could be detected in the removed solution using an atomic absorption spectrometer (AAS), and the dissolved Fe is given in table 1 expressed as ppm. These data suggest that CBD can dissolve not only ultrafine superparamagnetic (SP) magnetite, but also very coarse 355-500 μm magnetites. The amount of dissolved Fe (by AAS) was found to be inversely proportional to the grain size; the free iron dissolved in solution

is more than 30 times greater for the SD (single domain) than for the MD (multi-domain) magnetite. For the MD samples subjected to repeated CBD applications, it was found that after each successive CBD application, the amount of free Fe in solution was very similar, about 2300 ppm (Table 1), suggesting that for the coarser magnetites, the amount removed by the treatment is approximately constant, for up to five applications, and we might expect ultimately that the magnetites are completely destroyed. In fact Venegas et al. (1993) found that repeated CBD applications eventually removed all the free iron (aluminium-substituted haematite) present in some Argentinian soils.

The CBD technique was subsequently applied twice to two grams each of eight loess and palaeosol samples, four natural samples and four equivalent heated samples; the solutions from the two treatments were combined for AAS measurements. After treatment, unheated natural samples showed a green colour before drying; after drying, the powder still had a light green colour and all showed a large decrease in susceptibility, a similar result to that obtained by Verosub et al. (1993). Heated samples showed similar colours to their parent material before and after CBD treatment, and also a large decrease in susceptibility (Table 2). The AAS data suggest that the amount of Fe dissolved from both synthetic magnetite and loess (and palaeosols) is quite similar.

Thermomagnetic behaviour of synthetic magnetite and maghaemite

As indicated earlier, the magnetic characteristics of magnetite and maghaemite are quite similar, and this makes it difficult to separate them using magnetic techniques. However, thermal stability does provide a potential method of determining the maghaemite contribution to saturation remanence (Js or Ms) through the analysis of thermomagnetic (Js-T or Curie) curves.

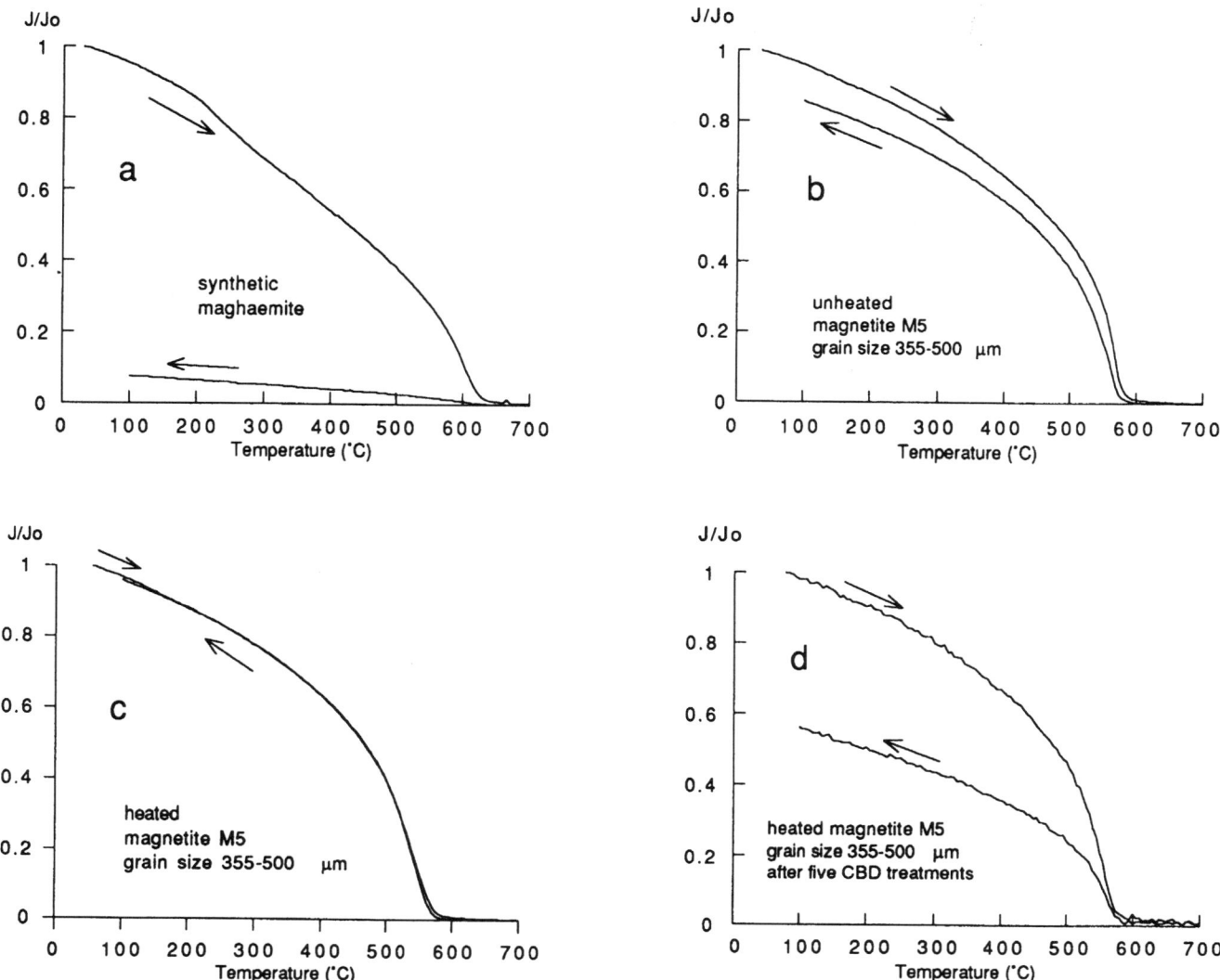

Figure 1. Thermomagnetic curves for synthetic ferrimagnetic powders. (a). maghaemite; (b) MD (355-500,μm) magnetite; (c) the magnetite sample after heating to 700 C°, and (d) the same heated sample after five applications of the CBD technique.

On heating, maghaemite is usually altered to haematite at temperatures above 300°C (Collinson 1983; Thompson & Oldfield 1986; cation-deficient (CD) magnetite goes to haematite and magnetite). In figure 1a we show Curie curves for a sample of synthetic maghaemite powder (J.K. Eyre, unpublished data; sample provided by B.A. Maher) dispersed in aluminium oxide. It can be seen that the synthetic maghaemite sample inverts completely on heating, the cooling curve indicating the presence of a single haematite phase. The Curie temperature (Tc) for the heating curve is about 630°C, close to a reported value (640°C; Özdemir *et al.* 1984) for maghaemite. This sample does not show a distinct change in the heating curve above 300 °C, but loses its magnetisation gradually until the Curie temperature.

A pure (stoichiometric) magnetite particle exposed to the air will gradually undergo low-temperature oxidation, producing a maghaemite (or CD magnetite) 'rim' on larger grains while the smallest grains may eventually become completely maghaemitised. The presence of maghaemitized material will be indicated by a Curie temperature between 575°C (magnetite) and 640°C (maghaemite), depending on the degree of oxidation, and will introduce irreversibility to the cooling curve, with a loss in room temperature magnetisation after heating. In Figure 1b, a thermomagnetic curve for 355-500 μm sized magnetite shows a degree of irreversibility which

we attribute to the initial presence of a partially-oxidised surface layer; a small degree of cation-deficiency is suggested by the Curie temperature for the heating curve which is some 8°C higher than the Curie temperature for the cooling curve. Repeating the thermomagnetic analysis produced fully reversible curves (Fig. 1c) indicating that a single heating was sufficient for the surface oxidised layer to invert to haematite. Figure 1d shows a thermomagnetic curve for the same sample after five CBD applications. The sample loses in excess of 40 % of its magnetisation during thermal cycling which implies that repeated CBD treatment has produced a thermally unstable ferrimagnetic phase.

Thermomagnetic behaviour of loess and palaeosols

The thermomagnetic behaviour of a loess (a; LCL9) and a palaeosol sample (b; LCl) are shown in figure 2 (a and b, solid lines). Both show a reduction in magnetisation after heating, this being a larger percentage for the loess sample (this result is consistent for sixty-four loess and palaeosol samples investigated in this way; the percentage reduction is always greater for loess). This loss of magnetisation is regarded as the main result of removing maghaemite from loess samples by heating (Heller *et al.* 1991; Liu *et al.* 1992; Maher & Thompson

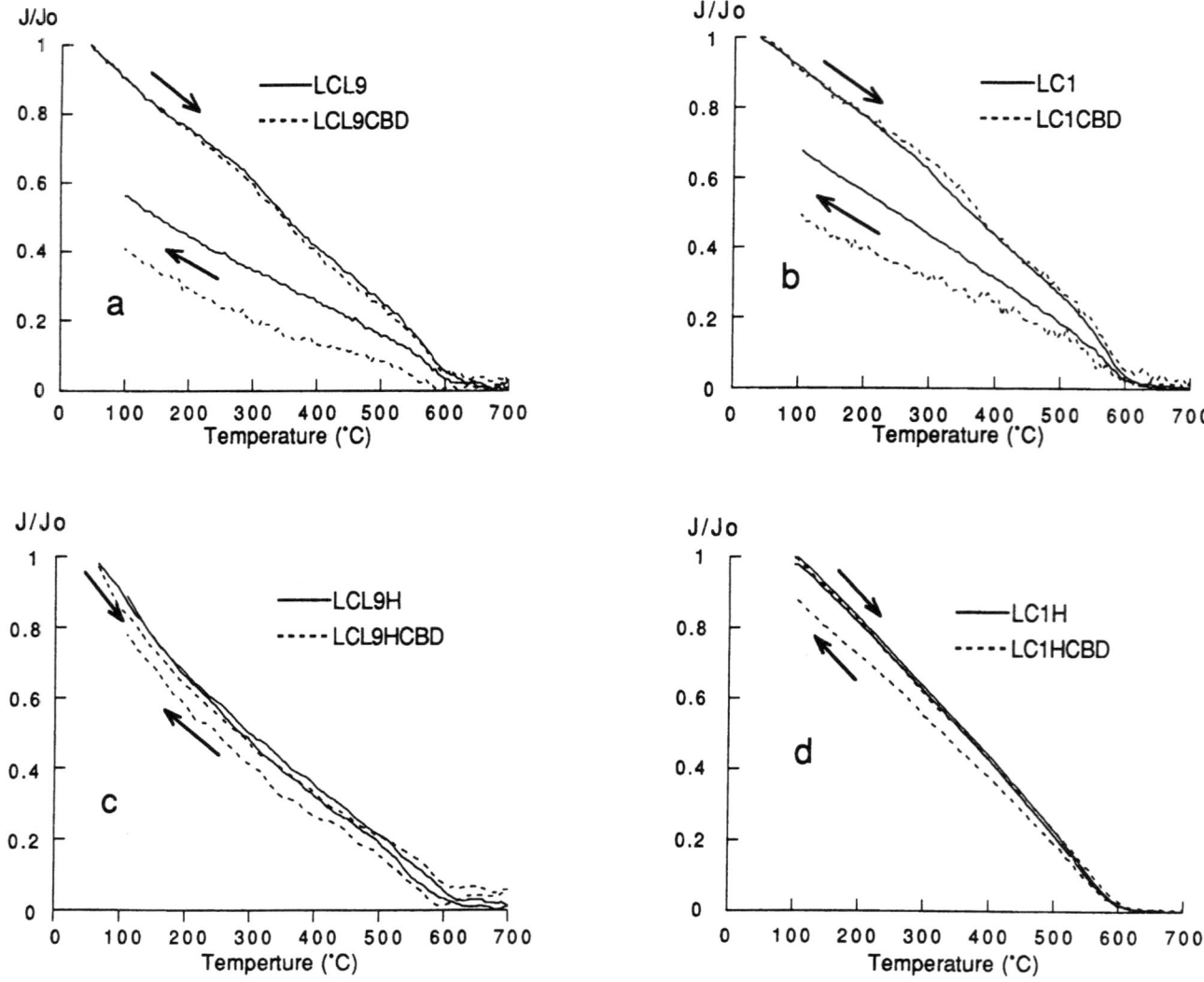

Figure 2. Thermomagnetic curves for loess and palaeosol samples before and after CBD treatment. a) unheated loess LCL9; b) unheated palaeosol LCl; c) heated loess LCL9 (LCL9H) and d) heated palaeosol LCl (LClH). All four samples show an increase in the magnetisation lost between heating and cooling, after the CBD treatment.

1992; Evans & Heller 1994). The Curie temperatures for the heating curves are close to 600°C, consistent with magnetite showing evidence of cation-deficiency (maghaemitization); the cooling curves have Curie temperatures close to 580°C, consistent with magnetite. Although we might expect part of this reduction in magnetisation to arise from the oxidation of ultrafine magnetite, this must only provide a minor contribution in view of the generally accepted predominance of such grains in palaeosols; if this fraction made an important contribution to the reduction in magnetisation after heating then we would expect the palaeosols to show the greatest reduction. These thermomagnetic results suggest to us that while maghaemite is important in both loess and palaeosols, it is by no means such a dominant fraction as the results of Verosub et al. (1993) seem to suggest (no more than 50% of Js in the samples we have studied). Furthermore, the implication of a larger percentage of maghaemite in loess than palaeosols is that the maghaemitized grains are a primary (aeolian) component and that the enhanced (pedogenic) magnetic fraction in the palaeosols is dominantly ultrafine magnetite.

To further investigate the conflicting results, we heated a selection of loess and palaeosol samples (to 700°C for one hour) to remove the maghaemite, and then subjected them (and the unheated parent sample) to a series of magnetic tests

and to the CBD extraction technique. The magnetic susceptibility and saturation magnetisation (Ms) values are given in Table 2. Figure 2 shows Curie curves for unheated and heated samples LCl and LCL9 taken from palaeosol S0 and loess L9 respectively (Luochuan section), both before (solid line) and after (broken line) the CBD treatment. The reversibility of the Curie curve for the heated samples (Fig. 2c, and 2d) indicates that the heating has successfully removed the maghaemite component. This is supported by the shape of the heating curves of the pre-heated samples which do not show the 'humps' (centred at 250°C) that are present in the unheated, natural samples (Fig. 2a and 2b).

After the CBD treatment, the thermomagnetic curves of all four samples are less reversible than before treatment, particularly the unheated samples. The increased loss of magnetisation on heating for the CBD treated samples is not only an increase in the relative magnetisation lost, but also in absolute terms. The influence of maghaemite on the heating curves between 250-400°C is apparent, and even enhanced for sample LC 1. These results provide further evidence that the CBD method does not remove maghaemite and, moreover, actually appears to enhance the thermally unstable magnetic component in these samples. After application of the CBD technique, all four samples show a large decrease in Ms (Table

Table 2 Magnetic measurements of loess samples before and after CBD

		pre-CBD			post-CBD		
Sample	**Location**	**Susc.** $(10^{-7} \, m^3kg^{-1})$	**Ms** $(10^{-2} \, Am^2kg^{-1})$	**Fe**** (ppm)	**Mass** (g)	**Susc.** $(10^{-7} \, m^3kg^{-1})$	**Ms** $(10^{-2} \, Am^2kg^{-1})$
LCl	Luochuan S0	11.84	6.70	2896	1.742	2.362	3.35
LClH	heated LCl	53.24	22.9	2558	1.886	22.07	1.15
LC4	Luochuan Sl	19.72	9.89	7695	1.664	2.032	2.50
LC4H	heated LC4	83.33	25.5	4752	1.38*	19.36	6.14
LCL9	Luochuan L9	2.03	2.23	3186	1.711	1.32	1.89
LCL9H	heated LCL9	1.97	1.03	402	1.808	0.89	0.87
BY61	Baicaoyuan Ll	3.205	3.86	2916	1.4946*	2.26	3.09
BY61H	heated BY61	36.48	5.29	2768	1.7956	24.1	4.88

*:some loss during treatment;   **: AAS measurements from CBD solution
Ms = Saturation magnetization

2), in agreement with the large decrease in magnetic susceptibility. As with the synthetic magnetites, the removed ferrimagnetic components can be found in the CBD solution using AAS (Table 2).

Discussion

A recent publication (Verosub *et al.* 1993) has considered the effect of using CBD on loess and palaeosols. They concluded that the CBD technique destroyed all the maghaemite and ultrafine magnetite, which they considered to be pedogenic, while leaving the (aeolian) magnetite untouched. On this basis they ascribed the dominant ferromagnetic fraction in the loess and palaeosols (up to 90%) to pedogenesis and suggested that this fraction was mostly maghaemite. Although the precise CBD chemical reaction is still not clear, our experiments with synthetic magnetite samples have shown that CBD can certainly dissolve magnetite covering a wide range of grain sizes. This result alone is sufficient to question the interpretation of Verosub *et al.* (1993) of the effects of the CBD method and thereby their inferences regarding the role of climate in the loess magnetic record. Furthermore, there is evidence that repeated CBD treatments actually produce thermally unstable ferrimagnets from these synthetic magnetite powders. After five CBD treatments, the 355-500 µm powder lost more than 40% of its magnetisation during thermomagnetic analysis. The heating curve does not show the 'hump' which is indicative of the presence of maghaemite, and this suggests that the CBD treatment has produced thermally unstable ultrafine magnetites which have oxidised upon heating. If this assumption is correct then we can infer that five CBD treatments is sufficient to reduce 40 % of the MD grains, in the range 355-500 µm, to a size small enough for oxidation to occur.

In the light of these results on synthetic ferrimagnetic material, our experiments on both heated and unheated loess and palaeosol samples have been an attempt to provide an insight into the performance of the CBD method on the Chinese loess. Our results for the unheated loess and palaeosol sample indicate that both contain a proportion of thermally unstable ferrimagnetic minerals and that after CBD treatment this proportion increases considerably. The Curie temperatures and behaviour of samples during thermomagnetic analysis suggests that this component is maghaemite/CD magnetite (with undoubtedly some contribution from SP ferrimagnets), which is altered to haematite upon heating. The increase in this unstable fraction after CBD treatment is an interesting result

and we can imagine a number of possible explanations. Firstly, the CBD treatment preferentially removes magnetite, thereby enhancing the contribution of maghaemite. However, the unstable fraction increases in absolute as well as relative terms, so this explanation cannot fully explain our observations. Secondly, the treatment produces a thermally unstable magnetite fraction, perhaps by shifting the magnetic grain-size distribution towards smaller grains (this may explain the result of the five CBD treatments on the synthetic MD magnetite powder). Thirdly, the CBD treatment produces new, thermally unstable ferrimagnetic material from the iron present in the silicate minerals or through the chemical alteration of existing stable ferrimagnets. A combination of the first and second options seems most likely and Figure 2a appears to confirm this conclusion. The shapes of the heating curves of LCL9, before and after CBD treatment, are almost identical, indicating that the CBD treatment has not removed the maghaemite fraction responsible for this characteristic shape. The cooling curves, however, whilst having the same shape, indicate that after CBD treatment the sample undergoes greater oxidation during the heating. This cannot indicate a larger maghaemite content for the second sample as both were prepared from an homogenised parent sample. Instead, this result indicates that not only does the CBD treatment fail to remove the maghaemite fraction, but it also produces a new thermally unstable ferrimagnetic phase. This behaviour can be explained in the context of the synthetic magnetite results. These indicated that while the CBD method is effective at dissolving coarse as well as fine-grained magnetite, it appears to be more effective with the SD size magnetite powder. If, as we believe, the maghaemite fraction resides in those magnetic grains of primary aeolian origin, and pedogenesis primarily produces ultrafine magnetite, then we might expect the CBD method to preferentially dissolve the magnetite fraction, removing the very finest fraction and at the same time reducing other grains to a size where they become thermally unstable.

Although this conclusion appears to support the idea that the CBD method can separate the detrital and pedogenic components, although not in the way suggested by Verosub *et al.* (1993), this would only be true if the aeolian magnetic input (and pedogenic component) were of a constant type and grain size (both in time and space) and if the efficiency of the CBD method was not related to the duration of the application. Clearly at any site the magnetic grain size of the aeolian input will be related to distance from source and will be modulated by such climatic parameters as, for example, wind velocity.

Therefore, considering aeolian input alone, we might expect the CBD technique to be most effective in the south-east of the loess plateau (source regions are towards the north-west) and during interglacials, when lower wind velocities will mean a finer aeolian input. A recent study (Guo *et al.* 1993) of the Xifeng loess section has suggested that up to two-thirds of soil thickness at the site can be attributed to aeolian input and that this aeolian input was richer in iron oxides than the aeolian input during loess forming periods.

Conclusion

We have found that contrary to suggestions from earlier studies, the CBD technique is very efficient at removing magnetite. Although it does show a grain size dependance, working with greater efficiency on single domain sized grains, it also removes a signficant fraction from large, multidomain grains. In addition, repeated applications of the technique removes approximately the same proportion, each time, from coarse magnetite grains, indicating a strong time dependence for the technique. When applied to loess and palaeosol samples, measurements of saturation magnetisation indicate that a single application of the technique removes between 20 and 70% of the ferrimagnetic content, with the larger fraction relating to palaeosols. Thermomagnetic analysis suggests that for the loess samples the technique leaves the maghaemite fraction unchanged, reflecting its primary (aeolian) and larger grain sized nature, while the finer grained fraction is preferentially dissolved. This finer grained fraction will be partly aeolian and partly pedogenic, the relative importance of each being related to both climatic conditions (glacial/interglacial) and distance from source area, as will the size of the fine fraction relative to the coarser fraction. Although in some parts of the loess plateau, especially during interglacials, this removed fine fraction will be dominantly pedogenic, the suggestion that the CBD technique can separate the pedogenic and primary components is an oversimplification both of the technique and the spatial variation of the magnetic grain size distribution and origin within the Chinese loess deposits.

Acknowledgements

We thank John Eyre for kindly providing synthetic maghaemite data, and Dr. W. Owens for providing synthetic magnetite samples. This research is supported by the Science and Engineering Research Council, UK (grant GR/H59268).

References

EVANS, M.E. & HELLER, F. (1994). Magnetic enhancement and palaeoclimate: study of a loess/palaeosol couplet across the Loess Plateau of China, *Geophysical Journal International* 117, 257-264.

COLLINSON, D.W. (1983). *Methods in rock magnetism and palaeomagnetism:* Chapman and Hall, London, pp. 503.

FINE, P. & SINGER, M.J. (1989). Contribution of ferrimagnetic minerals to oxalate- and dithionite-extractable iron: *Soil Science of America Journal*, 53, 191-196.

GUO Z.T., FEDOROFF N. AN Z.S. & LIU T.S. (1993). Interglacial dustfall and origin of iron oxides-hydroxides in the paleosols of the Xifeng loess section, China, *Scientia Geologica Sinica*, 2, 91-100.

HELLER, F., LIU X.M., LIU T.S. & XU T.C. (1991). Magnetic susceptibility of loess in China, *Earth and Planetary Science Letters*, 103, 301-310.

LIU, X.M., LIU T.S., HELLER, F. & XU, T.C. (1990). Frequency-dependent susceptibility of loess and Quaternary paleoclimate. *Quaternary Science*, 1, 41-50. (in Chinese with English abstract).

LIU, X.M., SHAW, J., LIU, T.S., HELLER, F. & YUAN, B.Y. (1992). Magnetic mineralogy of Chinese loess and its significance. *Geophysical Journal International*, 108, 301-308.

MAHER, B.A. & THOMPSON, R. (1992). Paleoclimatic significance of the mineral magnetic record of the Chinese loess and paleosols. *Quaternary Research*: 37, 155-170.

MEHRA, O.P. & JACKSON, M.L. (1960). Iron oxide removal from soil and clay by a dithionite-citrate system buffered with sodium bicarbonate: *Clay and Clay Minerals*, 7, 317-327.

ÖZDEMIR, Ö. & BANERJEE, S. K. (1984). High temperature stability of maghemite (γ-Fe$_2$O$_3$), *Geophysical Research Letters*, 11, 161-164.

ROLPH, T. C., SHAW, J., DERBYSHIRE, E. & WANG J. T. (1993). The magnetic mineralogy of a loess section near Lanzhou, China, In: Pye, K. (ed.), The dynamics and environmental context of aeolian sedimentary systems, *Geological Society Special Publication* 72, 311-323.

TAYLOR, R.M. & SCHWERTMANN, U. (1974). Maghemite in soils and its origin, I. Properties and observations on soil maghemites, *Clay Minerals*, 10, 289-298.

THOMPSON, R. & OLDFIELD, F. (1986). *Environmental magnetism.* Allen and Unwin, London.

VENEGAS, R., KANTER, F.L., ACEBAL, S., GRASSI, R., RUEDA, E.H., AGUIRRE, M.E. & SARAGOVI, C. (1994). Analysis of iron state in some Argentinian soils by dissolution methods and Mössbauer spectroscopy, *Hyperfine Interactions* 83, 451-455.

VEROSUB, K.L., FINE, P., SINGER, M.J. & TENPAS, J. (1993). Pedogenesis and palaeoclimate: interpretation of magnetic susceptibility of Chinese loess-palaeosol sequences, *Geology*, 21, 1011-1014.

Quaternary Proceedings No. 4, 1995 59-68
© Quaternary Research Association, Cambridge.

Soils in Aeolian Sequences as Evidence of Quaternary Climatic Change: Problems and Possible Solutions.

J.A. Catt

J.A. Catt, 1995 Soils in Aeolian sequences as evidence of Quaternary climatic change: problems and possible solutions, In *Wind Blown Sediments in the Quaternary Record* (Edward Derbyshire). Quaternary Proceedings No. 4, John Wiley & Sons Ltd., Chichester, pp. 59-68.

Abstract

Buried and surface soils in aeolian sequences hold great potential for Quaternary paleoclimatic interpretation, but this is limited at present by lack of knowledge of the complex relationships between soil properties and climatic factors and by our limited ability to estimate the lengths of soil-forming intervals. Polygenetic surface soils are likely to provide as much useful evidence of late Pleistocene climatic change as buried soils. A new definition of 'palaeosol' and a new system for classifying paleosols are required, to ensure better international communication in paleopedology and to clarify the effects of erosion and diagenesis. Micromorphological examination and mineralogical analyses help differentiate the often opposing effects of diagenesis and pedogenesis in buried soils. Most paleoclimatic interpretation of paleosols has been qualitative and should be regarded as tentative, but approaches to a quantitative interpretation based on the soil state factor model have also had limited success because of the difficulties in constructing reliable climofunctions. The best hope for the future is the construction and use of multivariate models of the soil-landscape continuum using geographical information systems to overlay comprehensive data sets of soil temporal and spatial variation.

KEYWORDS: soil, palaeosol, mineralogy, micromorphology, diagenesis, climofunction.

J.A. Catt, Soil Science Department, Rothamsted Experimental Station, Harpenden, Herts., AL5 2JQ, UK.

Introduction

Because aeolian sediments are deposited on land surfaces, they have often been subdivided using soils, which also form on land surfaces, though at times when there is little or no deposition. Soils are common in coversand and sandloess (or coverloam), but are more important in loess successions because in many parts of the world these span much longer periods (even the whole) of the Quaternary. It is usually possible to obtain information on past climates from these soils, but to do so reliably involves disentangling the various effects of time (length of soil-forming periods), other soil-forming factors and post-burial changes (diagenesis) on the soil properties. In this paper I discuss problems arising from these effects and suggest some methods for distinguishing them. However, it is first necessary to clarify some aspects of the definition, recognition, description and classification of soils that often cause misunderstandings. Most previous work on loess-soil successions has been concerned mainly with stratigraphical subdivision and dating; some simple proxy-climatic properties (*e.g.* magnetic susceptibility; silt:clay ratios) have been used to demonstrate correlations with the oxygen isotope record in deep ocean cores and other models of Quaternary climatic change, but this approach ignores a wealth of other information which might be obtained from loess sequences to give a more detailed record of Quaternary climatic change.

Definition and origins of soil

As the word 'soil' is used in many different senses by civil engineers, botanists, farmers, gardeners, etc., it is important to define it for use in Quaternary stratigraphy (Catt 1987, 1990, 1991). Most definitions by pedologists, geologists and geomorphologists emphasise processes of formation dependent upon proximity to the atmosphere. For example, at its 1993 inter-congress workshop, the INQUA Commission 6 (Palaeopedology) agreed the following definition: 'a three dimensional body on the surface of the earth composed of mineral and/or organic material, air and water, and formed by the impact of environmental factors acting on parent materials over a period of time to produce a sequence of horizons'.

For soils developed in unconsolidated sediments such as loess or coversand, the known processes (Duchaufour 1982; Birkeland 1984) include:

a) chemical weathering (*e.g.* decalcification; release of Fe or Al from primary minerals to form oxides or hydrated oxides),

b) downward leaching of soluble materials, including those released by weathering,

c) precipitation of soluble salts in near-surface horizons because of evaporation in an arid environment,

d) incorporation of decomposing organic material (humus),

e) disturbance by plant roots, burrowing animals, tree-fall, frost action or shrink-swell by drying and wetting of clays,

f) downward movement (eluviation) of soil particles, usually clay, in percolating water and redeposition (illuviation) in subsoil horizons,

g) reduction of brown, yellow or red ferric iron minerals to more soluble, grey ferrous compounds (gleying) in waterlogged conditions.

The relative importance of these processes varies from site to site according to changing parent material and environmental factors, thereby producing soil profiles with different sequences and thicknesses of horizons. Horizons are approximately parallel to the land surface and are differentiated according to chemical and mineralogical composition, particle size distribution, field properties such as colour (including mottling), strength, plasticity, stickiness, cementation and porosity (Soil Survey Staff 1951; Hodgson 1976) and microfabric characteristics visible in thin section (Bullock *et al.* 1985). Beneath land surfaces resulting from erosion, soil horizons are likely to be unconformable with rock structures such as bedding, but where land surfaces have been built up by deposition of sediment, as is usually the case with aeolian deposits, soil horizons are parallel to stratification and may coincide with lithostratigraphic units.

The definition of 'palaeosol'

According to Johnson and Hole (1994), the term 'palaeosol' was first used by Erhart (1932) for buried chernozem-like soils in the loess of Aachenheim, Alsace. Other buried soils recognised in the early decades of the 20th Century were usually termed 'fossil' soils (Latin: *fossus* = dug up), the earliest use of which term was probably by Ramann (1911, 1928). Ramann also recognised 'relict' soils, or surface (non-buried) soils which have lost some of their original properties because of a change in climate. Bryan and Albritton (1943) distinguished 'monogenetic' and 'polygenetic' soils, the latter corresponding to Ramann's 'relict' soils in having formed over two or more periods of different climate. Russian soil scientists such as Polynov (1927) and Nikiforoff (1943) had meanwhile recognised that the science of palaeopedology should include surface soils inherited from past periods when soil-forming conditions were different from those of the present (*i.e.* relict or polygenetic soils) as well as buried (fossil) soils. This is especially important if the properties of buried soils are used for palaeoclimatic interpretations based on comparisons with surface soils, which may or may not be in equilibrium with present-day climate.

Ruhe (1956) also realised the need to obtain palaeoclimatic and other evidence from these two types of ancient soils, and consequently defined 'palaeosol' as 'a soil formed on a landscape during the geologic past' (p. 441). Apart from encompassing soils of past periods that were either buried beneath younger sediments or have persisted at the surface through climate changes, this definition had the advantage of emphasising the fact that soils occur on landscapes (*i.e.* they extend over all parts of the land surface and vary in properties from place to place).

Although Ruhe's definition was supported and used by the INQUA Palaeopedology Commission (Yaalon 1971), it was later criticised by Bos and Sevink (1975), Catt (1987) and others because 'the geologic past' is inadequately defined. For example, does a soil which is suddenly buried by a landslip or lava flow locally become a palaeosol merely because of that event? At the site of burial the land surface and soil become past instead of present features, yet their clear correlatives beyond the limits of the landslip or lava flow are still present features. To overcome this problem, some palaeopedologists have placed a minimum age limit on palaeosols. Because climate is an important soil-forming factor, this limit is often placed at the Pleistocene-Holocene boundary, though Duchaufour (1982) defined palaeosols as soils formed before the last major cold period (Würm), and Catt (1979, 1989) has pointed out that climatic and other environmental changes in the mid-Holocene also influenced soil development in parts of Europe.

During and since its 1993 inter-congress workshop in Illinois, the INQUA Palaeopedology Commission has attempted to redefine 'palaeosol' to avoid these problems. However, New Zealand palaeopedologists have also raised the problem of soil-forming processes (*e.g.* mineral weathering) continuing for many thousands of years, possibly to the present day, in soils derived from and buried by easily weatherable volcanic deposits (tephras) in New Zealand. In regions some distance from the volcanic centres these soils form stacked sequences in successive thin tephras dating back to the Middle Pleistocene or possibly earlier. Do these buried soils occur on landscapes of the past if they are still influenced by processes related to the present land surface? Similar processes may also occur in stacked sequences of soils separated by thin loess layers. This may mean that 'palaeosol' (simply meaning 'old soil') will either have to be defined using some arbitrary age limit, such as the Pleistocene-Holocene boundary (10,000 radiocarbon years B.P.), or dropped in favour of a simple division into buried and surface soils.

Johnson and Hole (1994) and others have suggested that 'palaeosol' should be limited to the original definition of buried soil (Erhart 1932), and although this would avoid some of these problems, it contravenes current usage and is stratigraphically illogical in that it often divides soils which are demonstrably single units. For the moment a completely satisfactory definition seems especially elusive. Nevertheless, it is a useful word when used in a loose sense such as that of Ruhe (1956).

Identification of buried soils in aeolian sequences

The processes involved in soil formation outlined earlier bring changes in the colour, mineralogical and chemical composition, structure and fabric of aeolian sediments, which should enable buried soils to be identified and distinguished from sediment not modified in these ways. However, problems in the recognition of palaeosols in aeolian successions often arise because:

a) aeolian sediments are deposited on land surfaces and are subject to some pedogenesis even during deposition; the extent to which soils develop therefore depends principally on the rate of aeolian deposition, and the entire thickness of all true aeolian successions can (and should) be interpreted in terms of sequences of more or less strongly developed soils;

b) many weakly developed soil features can be difficult to distinguish from those resulting from changes in the detrital composition of the original sediment; this problem is most extreme with soil-sediments, which contain large amounts of transported but incompletely disaggregated soil material, and can sometimes strongly resemble the original *in situ* soil horizons;

c) diagenetic processes after burial can also produce

features resembling those in certain soil profiles; for example, groundwater can mobilise iron and manganese to produce areas depleted or enriched in these elements rather like gleyed soil horizons or podsol profiles; other diagenetic processes may mobilise and redeposit carbonates, or slowly remove humus originally incorporated during pedogenesis;

d) the friable upper horizons of many soils were often removed by erosion prior to burial beneath younger sediment, so that only the lower, less friable horizons persist; those are often also the less strongly modified horizons which provide less evidence for the nature and palaeoenvironmental implications of the original profile as a whole.

As Valentine and Dalrymple (1975,1976) suggested, these problems in identifying buried soils are most likely to be resolved by microscopic examination in thin sections, by tracing soils laterally to look for the characteristic changes that occur in relation to topography or other environmental factors, or by analysing a vertical sequence of samples to identify depth functions characteristic of soils, such as an upward decrease in weatherable minerals.

Classification of soils

Unfortunately there are many different (often national) systems of soil classification in use throughout the world, but no universally accepted system. For palaeosols, most palaeopedologists use whatever national system for surface soils that they have become used to, and this leads to considerable confusion and difficulties of correlation and interpretation. Probably the most widely used systems for surface soils are those of the United States Department of Agriculture (Soil Survey Staff 1975, 1992) and the Food and Agriculture Organization of the United Nations (FAO-UNESCO 1974,1988). The latter comes nearest to an international system in that it has been used for the legend for a world soil map, but its classes are less strictly and logically defined than those of the USDA and some other national systems. A particular problem of the USDA system, however, is its difficult terminology.

Most soil classification systems, though not that of FAO-UNESCO, are hierarchical in nature. The lowest level of classification is usually the soil series, which is a group of soils with similar character and arrangement of horizons in the profile and developed under similar conditions from one type of parent material. As originally applied in the USA, each series represented one or more areas of land and was named after an appropriate local town or village. However, to show relationships between series they are also grouped into higher categories mainly on the basis of properties reflecting broad genetic processes. In the USDA system the higher categories are known as great groups, suborders and orders, but these terms are not universal; for example, in the classification for England and Wales (Avery 1980) they are soil subgroups, soil groups and major soil groups, usually with different differentiating criteria from the American higher categories. The higher categories are sometimes used for map units at smaller scales (e.g. 1:1 000 000) than those appropriate for soil series (1:25 000 - 1:100 000).

Almost all systems for classifying surface soils were designed with practical applications especially to agriculture in mind. This limits their value in palaeopedology. In the American system, for example, present climatic factors (rainfall and temperature) are used at a fairly high level to distinguish classes; this is obviously useful as an indicator of potential plant growth, but introduces confusion with respect to polygenetic soils, many of the properties of which originated in past periods when the climate was different from that of the present (Bronger & Catt 1989). Another problem in using these classification systems for palaeosols is their dependence on the properties of surface horizons, which have often been lost from buried soils because of erosion (Bronger 1980).

The INQUA Palaeopedology Commission has been developing a classification system for palaeosols based on genetic processes, but at present there is little agreement on this. Until this is available, the various systems currently used will continue. However, more important than putting a possibly ill-defined name to a buried soil is that field and laboratory properties should be described in as much detail as possible so that processes of origin can be inferred. Where buried soils are related to surface soil types it should be at the lowest possible hierarchical level in a stated classification system (e.g. great groups in the USDA system), and the limitations of any such affiliation should be discussed fully and clearly.

Horizon designations

Soil horizons are commonly given interpretative symbols, which are based upon morphology and implied genesis and often provide information about neighbouring horizons above or below in the soil profile (Bridges 1990). Although there is a greater measure of international agreement over horizon designations than over soil classification, there is still some variation in the designations used.

Originally Russian pedologists such as Dokuchaev used the A, B, C system for simple labelling of horizons from the top downwards in chernozem profiles. Problems arose when these were applied to other soil types such as podsols, and Ramann (1911) proposed that A should be used for surface horizons enriched in humus and from which materials have been lost by leaching and illuviation, that B should be used for weathered subsoil horizons enriched in the illuvial materials, and C for lower horizons in which no effects of chemical weathering can be distinguished in the field but physical disruption has occurred. The USDA followed this system but added subscript numbers for separable A or B horizons (B_1, B_2, B_3), subscript ca and cs for C horizons containing accumulations of calcium carbonate and calcium sulphate, respectively, G for intensely gleyed horizons, and D for any stratum underlying and different from C horizons from which the remainder of the profile had developed (Soil Survey Staff 1951). A soil horizon nomenclature working group of the International Soil Science Society later introduced the E (eluvial) horizon, which is pale-coloured and has lost material by eluviation, and thus restricts the A horizon to one of humus incorporation. A 1962 supplement to Soil Survey Staff (1951) introduced O for horizons containing >30% organic matter (if mineral fraction is > 50% clay) or > 20% organic matter (with no clay).

Soil Survey Staff (1951), Kubiëna (1953), Whiteside (1959) and others recognised the value of postscript letters to indicate one or more processes active in the formation of a horizon. Various sets of these have been proposed. The most recent is that currently used by USDA (Soil Survey Staff 1992), and those useful with palaeosols are:

a: very decomposed organic material (< 17% fibre)
b: buried horizon
c: accumulation of concretions or nodules of Fe, Al, Mn

or Ti compounds

e: weakly decomposed organic material (> 17% fibre)

g: gleying (chroma of 2 or less on Munsell Color Chart resulting from waterlogging)

h: illuvial accumulation of humus in B horizon

k: accumulation of Ca (or other alkaline earth) carbonates (replaces ca)

m: continuous or nearly continuous cementation

n: accumulation of exchangeable Na

o: residual accumulation of sesquioxides

p: disturbance of A or O horizons by ploughing or similar other human-induced disturbance

q: accumulation of secondary Si

r: weathered or soft bedrock (C horizon)

s: illuvial accumulation of sesquioxides in B horizon, often complexed with humus

ss: presence of 'slickensides' resulting from swelling of clay minerals and shear failure

t: accumulation of silicate clay either formed within or eluviated into a B horizon

v: presence of Fe-rich, humus-poor red material that hardens irreversibly when exposed to the atmosphere (plinthite)

w: a weathered horizon showing development of pedological colour or structure but little or no eluvial accumulation

x: a firm, brittle dense horizon, often with prismatic structure and bleached vertical ped faces

y: accumulation of gypsum (replaces cs)

z: accumulation of salts more soluble than gypsum

Diagenesis in buried soils

If a soil is buried beneath a thin deposit and pedogenesis recommences for a significant period in this overlying deposit, the later soil may extend downwards and is eventually 'welded' (Ruhe & Olson 1980) to the earlier soil to form a pedocomplex. However, with deeper burial this is impossible, and the soil is likely to be modified by new (diagenetic) processes. In soils derived from and buried by aeolian deposits, which are usually quite permeable, percolating water often reverses the effects of pedogenesis, for example redepositing carbonate in horizons previously decalcified, increasing the pH or base saturation, or gradually removing the less resistant humic components of A horizons, mainly by bacterial decomposition. Because of this it is often impossible to place buried soils in classification systems for surface soils that use these properties as important differentiating criteria. Special techniques must be used to distinguish the diagenetic and earlier pedogenetic effects; foremost among these are thin section study (micromorphology) and mineralogical analyses.

Diagenetically redeposited carbonate derived from younger soils above may be difficult or impossible to distinguish from the pedogenetic secondary carbonate formed in deep subsoil (Ck) horizons. In thin section both may occur as subspherical micritic concretions, as coarser sparry crystals in larger voids, such as root channels, or as fine acicular crystals (lublinite) in small pores; however, different generations may occur in different forms or show other micromorphological characteristics, and can be distinguished from primary (detrital) carbonate which usually occurs as individual grains dispersed in the sediment matrix. Humic A (Ah) horizons often have an open spongy fabric in thin section, which may be preserved in buried soils even after removal of much of the organic matter. However, with very deep burial the open fabric may be partially closed by compaction, especially in clay- or humus-rich horizons.

Detrital and many pedogenic minerals are unaffected by the mild diagenesis to which most Quaternary buried soils are subjected. However, compaction and an increase in temperature with deeper burial favour dehydration processes, which contrast with the hydration reactions typical of chemical weathering in soils. For example, hydrated iron oxides such as goethite and ferrihydrite may be partially dehydrated to haematite, a process that may change the colour of the soil slightly (Walker 1967; Blodgett 1988). Alternatively, if an originally well-drained soil is influenced by a groundwater table after burial, brown or red iron oxides may be locally reduced to grey ferrous minerals, particularly where they are associated with root or other organic remains. Diagenesis can also bring changes in alumino-silicate clay minerals, though these are rarely as extensive as the better known diagenetic effects in the more reactive marine sediments (Larsen & Chilangar 1967). Initial changes occur in the adsorbed cations by exchange with groundwater containing other ions, such as Ca^{2+}, Mg^{2+} or K^+, but with increasing time and depth of burial there may be changes in interlayer ions, such as replacement of interlayer water molecules or the sodium, calcium or organic ions in smectites by K^+, leading to formation of mica-like layers with smaller interlayer spacings. Deeper burial may lead to further dehydration processes, such as the conversion of gypsum to anhydrite. Other changes in soils after deep burial are discussed by Retallack (1990) and Catt (1990).

Quantitative climatic evidence from soils

Early interdisciplinary work on the origin and classification of surface soils by Russian workers such as V.V. Dokuchaev and K.D. Glinka and by Americans such as E.W. Hilgard and C.F. Marbut (Tandarich & Sprecher 1994) showed that soil characters are influenced by a wide range of factors. This was expressed by Jenny (1941, 1980) in the factorial model:

$$S = f(cl, o, r, p, t, \dots)$$

in which S stands for soil, cl for climate, o for organisms, r for relief, p for parent material, t for time, and the dots indicate possible additional factors. These were not conceived as causes or forces, but as factors that explain the 'state and history of a soil' (Jenny 1980, p. xi). Nevertheless, Jenny believed that the equation could be solved given ideal conditions, in particular that individual factors can be evaluated in a sequence of soils in which all the others are held constant. For example, he proposed that functional relationships between S and cl (i.e. between individual soil properties and individual climatic factors, such as mean annual rainfall or mean annual temperature) can be established in sequences of surface soil profiles formed under uniform conditions of the remaining factors (o, r, p, t, ...). Such soil sequences are termed climosequences and the mathematical (or graphical) relationships established from them are known as climofunctions.

Using early analytical data for loess-derived surface soils of the virgin prairie grasslands of mid-western USA, Jenny (1941) constructed climofunctions for topsoil nitrogen content and mean annual rainfall, and for topsoil clay content and mean annual rainfall. These were possible because the original organisms (prairie grassland), relief (level or gently sloping plains), parent material (loess) and time (period of soil development since deposition of the loess in the late

Wisconsinan) were all reasonably uniform, though some of the soils were on Bignell Loess and others on Peoria Loess, and some were forested for part of their history and others were not (Ruhe 1984). Jenny also integrated the temperature and rainfall data into a nitrogen-climate surface or three-dimensional climofunction. Later climofunctions have also been established mainly for topsoil organic matter, carbon, nitrogen, C/N, and base saturation on a range of parent materials (Yaalon 1975). However, these relationships are rarely useful with buried soils because the properties involved are easily changed on burial or are the first to be lost through erosion.

Climofunctions for more permanent soil properties, such as SiO_2/Al_2O_3, SiO_2/Fe_2O_3 or Fe_2O_3/Al_2O_3 ratios in the subsoil clay fraction or total soil, have been established principally in soils derived from crystalline igneous rocks, in which it can be assumed that the parent material contained no free Al_2O_3 or Fe_2O_3. These also cannot be used with soils derived from aeolian or most other sediments, as the original Al_2O_3 or Fe_2O_3 content of these is uncertain.

Perhaps the most helpful climofunction is the relationship between depth to calcic (Ck) horizon and mean annual rainfall. This was originally investigated by Jenny and Leonard (1935) in loess-derived soils of the Great Plains. Further data for drier areas in the Mojave Desert were added by Arkley (1963) and Marion et al. (1985) and for Israeli soils by Dan and Yaalon (1982). The last also included data on depth to gypsum (Cy horizon) and the presence of more soluble salts. Retallack (1994) re-examined the relationship for carbonate in 317 freely-drained Holocene or late Pleistocene profiles on unconsolidated sediments (excluding clay and limestone) on low rolling terrain. As in the earlier work, the relationship is not linear but curved (Fig. 1), with a best fit expressed by D = -40.49 - 0.0852 P - 0.0002455 P² (correlation coefficient r = 0.78, standard error s = ± 33 mm), where D is the depth (cm) to the top of the Ck horizon and P is the mean annual precipitation (mm). As Retallack (1994, p. 38) pointed out, this result cannot be used to estimate the rainfall when a buried soil was formed if the profile was truncated before burial. However, the shape of the curve means that truncation causes a smaller error in the estimation of rainfall if the soil formed under moderately wet conditions (600 - 1000 mm rainfall). The gently sloping part of the curve (smaller rate of increase in D) occurs between P = 100 mm and approximately 800 mm, and in this range loss of 50 cm by erosion decreases the estimated value of P by 150-300 mm. Loss of 50 cm from a profile formed under a rainfall of 800-1000 mm results in a maximum error of only 120 mm.

Since publication of Jenny (1941) many pedologists have drawn attention to limitations of the soil state factor model (Smeck et al. 1983; Birkeland 1984). Although Jenny recognised the possible existence of other factors, they were not discussed; many, such as fires and influxes of particulate or dissolved material from aeolian or slopewash sources, are localised or sporadic, and it is difficult to allow for their effects in individual profiles. Although Jenny also recognised that some of the factors are inter-related (1941, p. 16), the model has often been criticised because the factors are not completely independent variables; for example, climate and relief are dependent variables in that temperature decreases and rainfall often increases with increasing altitude. Factor interactions are likely to be greater in older soils. Also the older a surface soil, the more likely it is to be polygenetic and thus have resulted from two or more different combinations of factors; the best climofunctions are derived from recent (e.g. Holocene) climosequences, and their extension to older surface soils and landforms is often impossible.

Later work also demonstrated that, even where soils should be uniform according to the state factors, they are very variable. The coefficients of variation of many soil properties exceed 25% over distances of only a few metres (Wilding & Drees 1983), and this seriously decreases the precision of climofunctions. Original lateral and vertical variations in the composition of many parent materials are an important source of this variability, and where further differences have been superimposed by weathering and other pedogenetic processes it can be very difficult to establish the exact nature of the parent material of many soil horizons, even by detailed petrographic studies.

Jenny included human influences in the o (organisms) factor, but in many soils these have had unknown but probably quite large effects that could confuse earlier relationships between climatic factors and soil properties. Liming, application of fertilisers or organic manures, disturbance by ploughing or

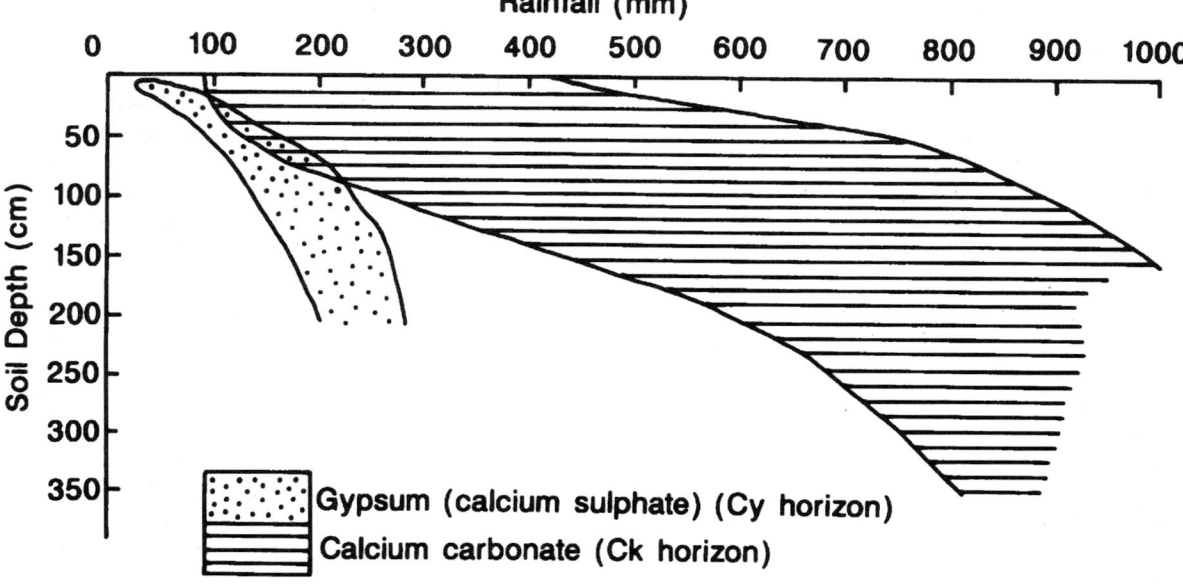

Figure 1. Relationship between depth to Ck and Cy horizons and mean annual precipitation; from Dan and Yaalon (1982) modified for Ck horizons using data from Retallack (1994).

subsoiling, irrigation, under-drainage and other soil treatments may all affect soil properties (pH, organic matter content, etc.) that in a purely natural environment would probably show clear relationships to climatic factors. As pointed out by Johnson and Hole (1994), the *o* factor should also include biomechanical processes such as soil disturbance by tree fall and soil fauna, but these were not considered by Jenny or subsequent workers; almost all *o*-factor evaluations have been limited to the chemical effects of vegetation types on soil.

A further limitation of the soil state factor approach is that some soil properties can themselves develop to a point where they change soil development processes independently of exogenous environmental factors such as climate. An example which commonly affects soils developed in loess is deposition of illuvial day in the pores and channels of subsoil (Bt) horizons to the extent that movement of water becomes restricted and the processes of gleying are initiated in and sometimes above the Bt horizon. As Johnson *et al.* (1990) noted, this type of pedogenetic change unrelated to soil state factors complicates not only the construction of functional relationships such as climofunctions but also the distinction between monogenetic and polygenetic soils.

Despite its failings, many of which were recognised by Jenny (1941, 1980), the factorial/functional model of soil formation has stimulated much detailed work on soil properties in relation to environmental factors and on the development of new pedogenetic models, such as the energy flux model (Runge 1973), the chemical equilibrium residual model (Chesworth 1973), the soil-landscape systems model (Huggett 1975), the progressive-regressive evolution model (Johnson *et al.* 1990), the systems mass-balance process model (Chadwick *et al.* 1990) and various simulation models (Levine & Ciolkosz 1986; Bryant & Olson 1987; Hoosbeek & Bryant 1992). Also a comprehensive four-dimensional approach for modelling the soil-landscape continuum, utilising all available field data, spatial analysis and geographical information systems has recently been proposed by McSweeney *et al.* (1994). This would include investigations of more complex interactions between soil properties and climatic factors using multivariate statistics rather than the bivariate or trivariate relationships established by Jenny (1941).

These models attempt in various ways to account for all soil-forming factors and processes and their interactions. For many buried soils, especially those not truncated by erosion and in which the effects of diagenesis can be assessed, use of models such as that of McSweeney *et al.* (1994) holds considerable promise for quantitative reconstruction of past climates. They would require much detailed information, especially on lateral variation of a soil in relation to parent material and relief, but this is essential to achieve the most reliable results. In addition to using buried soils in this way, it would be wrong to ignore the palaeoclimatic evidence which may be obtained from polygenetic surface soils developed in datable aeolian sediments. With these, soil development modelling should be able to provide reliable quantitative palaeoclimatic data for at least the Holocene, and possibly for much of the last 100,000-150,000 years where the surface soils have developed beneath stable land surfaces on older (*e.g.* oxygen isotope stage 6) loesses.

Qualitative climatic evidence from soils

Because of the difficulties involved in quantitative assessment of the effects of climatic factors on soil properties, almost all palaeoclimatic interpretation of buried or surface palaeosols

has been qualitative or semi-quantitative. This has mainly been based upon the early Russian concept of soil zonality (Ramann 1911,1928), that the main climate-vegetation zones of the earth are characterised by different soil types. This is a great over-simplification, as it ignores the other, possibly more localised, effects of the *r,p,t* and other factors. However, it does have sufficient validity to be useful for tentative general interpretation of palaeosols, again provided truncation and diagenesis are allowed for. Soil properties of significance, such as B or C horizons enriched in illuvial clay, carbonates, sulphates, other soluble salts, sesquioxides, etc., the mineralogical composition of clay fractions, reddening (rubefication) related to forrnation of haematite and the types and vertical distribution of humus are discussed in detail by Yaalon (1971), Birkeland (1984), Catt (1988a, 1991), Retallack (1990) and others.

However, the complex relationships between these soil properties and the range of soil-forming factors complicates most interpretations. For example, Bt horizons are commonly preserved in buried soils in loess sequences, and are usually taken to indicate an interglacial climate warm and moist enough to support deciduous forest vegetation. From its occurrence in surface soils and from experimental evidence, McKeague (1983) showed that clay eluviation can in fact occur in a wide range of situations. It is favoured by a pH of 4.5-6.5, or a higher pH if this is associated with abundant exchangeable sodium (as in some arid regions), by small amounts of cementing and flocculating agents (carbonates, humus, sesquioxides, exchangeable Al, Mg, Ca), by a system of fissures or channels such as those formed by roots or by dissolution of carbonate clasts, and by a seasonal distribution of rainfall. Many of these factors are consistent with forest vegetation but are not confined to it. If the structure of A and E horizons is stable, the particles illuviated into Bt horizons are essentially fine layer silicate clays with subsidiary but variable amounts of iron oxides, but as the soil structure weakens, coarse clay, fine silt and even humus may accompany the fine clay. This may occur under prolonged cultivation or with decreasing temperature, both of which weaken inter-particle bonds by loss of humus. Eluviation may therefore occur in some soils affected by seasonal freezing (McKeague *et al.* 1973); the high dielectric properties of meltwater disperse particles, which may then be moved downwards, coarser particles being redeposited in fissures in higher parts of the profile and clay at greater depths (Van Vliet-Lanoë 1985).

The different origins of clay accumulations in Bt horizons can often be distinguished in thin section from their optical properties (dustiness, colour, interference colour patterns) and their relationships to voids. Sometimes it is also possible to infer climatic changes with time from the micro-stratigraphy of the coatings or from their disruption by frost action in cold stages (Kemp 1987).

In very cold conditions most of the chemical processes of soil development dependent upon water (chemical weathering, leaching, humus incorporation, eluviation, gleying) ceased or were much less effective because of deep continuous freezing, and arctic structure soils (Mückenhausen 1982) were formed instead. Structural features in soils, such as ice-wedge casts, cryoturbations, vertically orientated stones, fragipans and various types of patterned ground, provide good evidence in mid-latitude regions for episodes of periglacial climate, and some form only below certain sub-zero temperature thresholds (Washburn 1979). The effects of frost are seen more commonly in thin section; they include disruption of earlier illuvial coatings and their incorporation into the soil matrix, fragmentation of plant tissue by needle ice, rounding of granular aggregates, and features resulting from illuviation by meltwater, such as silty cappings on stones or coarse sand particles, thin dense bands giving a platy microstructure and flat silty-clay droplet structures several cm across (Mücher & Morozova 1983). More experimental work could be done to determine the exact temperature and other conditions under which these form. Because loess and other aeolian sediments are often deposited on land surfaces in periglacial conditions, many such features are likely to occur in parts of the successions between the more obvious interglacial and interstadial soils, and should be studied, especially in thin section, for the greater detail of palaeoclimatic history they can provide (Huijzer 1993).

A further source of palaeoclimatic evidence from soils is provided by measurements of stable isotope ratios in pedogenic carbonates (Cerling 1984; Cerling & Hay 1986; Amundson et al. 1989). This could be extended to the buried Ck horizons in loess sequences, in particular to help distinguish primary detrital, secondary pedogenetic and diagenetic carbonates.

The effect of time

Most soil-forming processes are slow and produce visible effects (formation of recognisable soil profiles) only over fairly long periods of time; in some instances this can occur within tens of years, but many soils take centuries or thousands of years to develop. Any buried soil therefore provides evidence for a significant period of landscape stability, without either deposition to prevent soil development or erosion to remove a profile already formed. It follows that a soil represents a period of time rather than a point in time, though the imprecision of current dating methods may mean that some older, weakly developed palaeosols may have to be seen as points in time rather than periods.

However, small amounts of either erosion or deposition can occur during a period of soil development without necessarily removing the evidence for pedogenesis completely. If deposition is slow or occurs in small intermittent increments, the sediment is altered to some extent (e.g. by incorporation of humus) and the soil grows upwards by development of an over-thickened A horizon, the lower parts of which may later acquire some characteristics of a B horizon. These accretionary soils are common in aeolian successions; for example, where buried humic A horizons are overlain by a later sediment, the overthickened A horizon often passes gradually upwards into unaltered sediment, indicating that the rate of deposition was increasing and overtaking the rate of humus incorporation. In contrast, slow continuous or intermittent erosion during soil development usually results in a thin weakly developed profile, often lacking a B horizon (i.e. with a humic A horizon directly over an unaltered or little altered C horizon). This may lead to incorrect climatic interpretation because buried AC profiles are usually interpreted as weakly developed soils representing short or cool intervals (e.g. interstadials).

Buried soils consisting of a well-developed B horizon sharply overlain by fresh unaltered sediment indicate an episode of rapid erosion following a prolonged period of landscape stability. These are also common in aeolian sequences, and probably indicate a rapid climatic change to colder, drier conditions with disruption of the earlier vegetation cover and loss of friable topsoil by wind erosion, runoff or gelifluction before deposition of the fresh, overlying periglacial aeolian sediment under even drier conditions.

One of the main problems in palaeoclimatic interpretation of soils is separating the often reinforcing effects of time and climate on soil properties. The extent to which many properties (e.g. profile depth, rubefication, alteration of clay or other primary minerals, thickness of Ck or Cy horizons, illuvial clay content of B horizons) are developed depends upon the time profiles have taken to form as well the climate during those periods. To estimate the climatic factors one must first estimate the length of the relevant soil-forming interval. However, this is not easy. By reorganising the factorial model (Jenny 1941) to isolate the effect of time, it is possible to develop relationships between soil properties and time (chronofunctions, soil development indices) in profiles developed on similar parent materials and in uniform environmental conditions (Yaalon 1975; Harden 1982), but these cannot then be used to estimate soil-forming intervals in profiles developed in unknown climatic conditions. A soil's stratigraphic relationship to dated overlying sediments and parent materials provides an estimate of the maximum length of the soil-forming interval, but this may be much greater than the actual interval if the soil developed on an erosion surface (Catt 1988a, Fig. 3.8). Many soils in periglacial aeolian successions correspond to the time-transgressive chronosequences without historical overlap of Vreeken (1975), and in these the ages of parent materials and sediments overlying the soils probably provide much better estimates of the soil-forming intervals, because the deposits accumulated almost throughout successive cold stages of the Quaternary, which have been fairly precisely dated in oceanic sediment cores (Catt 1988b).

Bockheim (1980) investigated the relationship between time and climatic factors in 32 well-dated chronosequences from seven climatic regions (tropical rainy to cold desert); various parent materials were involved but all sites were on level or gently sloping landforms. He found that the best relationship between certain soil properties (depth of oxidation, profile thickness, electrical conductivity of salt-enriched horizons, total N and bulk density of the surface mineral horizon, clay content of B horizon) as a group (y) and time (x) was the logarithmic function: $y = a + b \,(\log x)$, in which $\underline{b}$ represented climatic factors and a was a constant with a ratio to b of 20:1 for most soil properties. For most of these properties b was mean annual temperature, but for bulk density it was mean annual precipitation. The equation could predict the soil property-age relationships over short (<500 yr) and long (>10^6 yr) periods, and because it is a logarithmic function it agrees with the suggestion of Yaalon (1971) and others that many soil-forming processes are initially rapid but become progressively slower as the soil matures. Of these properties the ones most likely to survive burial were depth of oxidation, profile thickness and clay content (illuvial plus weathering) of the B horizon, all of which would indicate mean annual temperature provided the length of the soil-forming interval is known.

Only two of the soil properties studied by Bockheim showed similar logarithmic relationships to time regardless of differences in climate and other soil factors; these were pH and base saturation of the surface mineral horizon. These properties are likely to provide good chronofunctions applicable to soils in all environments but little or no evidence of climate, and as they change rapidly during soil development and also after burial they are likely to have little value with buried soils.

Using the logarithmic function of Bockheim (1980) to distinguish the effects of time (lengths of interglacial periods indicated by the dating of oceanic sediment cores) and climate, Catt (1988b) suggested a worldwide climatic ranking of interglacials (oceanic oxygen isotope stages 1-23) based on the properties of soils in the main loess regions. The properties used had to be inferred from partial descriptions or local classifications of the soils attributed to each warm stage, and the ranking consequently included the combined effects of temperature and precipitation. A more precise comparison of interglacial temperatures might have been obtained if the calculations had been based on the three soil properties most closely related to mean annual temperature and most likely to survive burial (depth of oxidation, profile thickness and clay content of B horizon). Unfortunately, few descriptions of buried soils have included measurements of these properties, though they are all quite easily obtained from simple profile descriptions and examination of thin sections. This emphasises the importance of routinely recording simple soil properties, especially Munsell colours and thicknesses of different horizon types, and of taking undisturbed samples for preparation of thin sections whenever buried soils are exposed or sampled for other purposes (*e.g.* magnetic susceptibility measurements).

Conclusions

Buried and surface soils in aeolian sequences have considerable potential for providing a detailed history of Quaternary climatic change in many parts of the world. This could be more detailed than the records currently obtained from deep sea cores or thick lacustrine sequences and longer than those from ice cores. However, the full potential is difficult to realise at present because of the complex relationships between soil properties and climatic factors, the reinforcement of soil-climate relationships by length of soil-forming intervals (themselves difficult to estimate), and the imprecisely known effects of other soil-forming factors (parent materials, organisms, relief, etc.). The quantitative climofunction approach has consequently failed to supplant qualitative interpretation of soil profile types based mainly on the incomplete zonal concept of surface soil variation.

Some progress in separating the effects of climate (principally mean annual temperature) and time has been made by studying selected soil properties in well-dated chronosequences of surface soils in different climatic regions. However, the simple properties which would allow these relationships to be used for quantitative palaeo-temperature interpretation of buried soils have rarely been recorded. For the future the main hope of better palaeoclimatic interpretation is the development of multivariate models for the soil-landscape continuum using geographical information systems. However, the temporal and spatial analyses involved in these models will require large datasets of properties measured in the field, in thin section or in the laboratory. Past deficiencies in the description of buried soils should be remedied now, so that these datasets are already available when the models are ready for use.

At present much of the detailed palaeoclimatic information obtainable from the wide range of pedological features existing throughout all aeolian sequences is simply being ignored. The soils being studied are generally the more spectacular interglacial profiles, and intervening loess layers are often incorrectly assumed to show no pedological features of palaeoclimatic significance. Even where there is no macroscopic evidence of soil development in aeolian sequences, it is worth looking for the microfabric, chemical or mineralogical features that indicate the various types of weaker pedological reorganisation which would have occurred in cooler (stadial and interstadial) conditions. This enables a more detailed and more complete palaeoclimatic record to be reconstructed from aeolian sequences.

References

AMUNDSON, R.G., CHADWICK, O.A., SOWERS, J.A. & DONER, E.R. (1989). The stable isotope chemistry of pedogenic carbonate at Kyle Canyon, Nevada. *Soil Science Society of America Journal*, 53: 201-210.

ARKLEY, R.J. (1963). Calculation of carbonate and water movement in soil from climatic data. *Soil Science*, 96: 239-248.

AVERY, B.W. (1980). Soil classification for England and Wales (Higher Categories). *Soil Survey Technical Monograph 14*, Rothamsted Experimental Station, Harpenden, 67 pp.

BIRKELAND, P.W. (1984). *Soils and geomorphology*. Oxford University Press, New York, 372 pp.

BLODGETT, R.H. (1988). Calcareous paleosols in the Triassic Dolores Formation, Southwestern Colorado. *In:* J. Reinhardt and W.R. Sigleo (Editors), *Palaeosols and weathering through geological time: principles and applications*. Geological Society of America Special Paper 216, pp. 103-121.

BOCKHEIM, J.G. (1980). Solution and use of chronofunctions in studying soil development. *Geoderma*, 24: 71-85.

BOS, R.H.G. & SEVINK, J. (1975). Introduction of gradational and pedomorphic features in descriptions of soils. A discussion of the soil horizon concept with special reference to palaeosols. *Journal of Soil Science*, 26: 223-233.

BRIDGES, E.M. (1990). Soil horizon designations. *International Soil Reference and Information Centre Technical Paper* 19.

BRONGER, A. (1980). Zur neuen "Soil Taxonomy" der USA aus bodengeographischer Sicht. *Petermanns Geographisches Mitteilungen*, 124: 253-262.

BRONGER, A. & CATT, J.A. (1989). Paleosols: problems of definition, recognition and interpretation. *Catena Supplement* 16: 1-7.

BRYAN, K. & ALBRITTON, C.C. (1943). Soil phenomena as evidence of climatic changes. *American Journal of Science*: 241, 469-490.

BRYANT, R.B. & OLSON, C.G. (1987). Soil genesis: opportunities and new directions for research. *In:* L.L. Boersma (Editor), *Future developments in soil science research*. Soil Science Society of America, Madison, pp. 301-311.

BULLOCK, P., FEDOROFF, N., JONGERIUS, A., STOOPS, G. & TURSINA, T. (1985). *Handbook for soil thin section description*. Waine Research Publications, Wolverhampton, 152 pp.

CATT, J.A. (1979). Soils and Quaternary geology in Britain. *Journal of Soil Science*, 30: 607-642.

CATT, J.A. (1987). Palaeosols. *Progress in Physical Geography*, 11: 487-510.

CATT, J.A. (1988a). *Quaternary geology for scientists and engineers*. E. Horwood, Chichester, 340 pp.

CATT, J.A. (1988b). Soils of the Plio-Pleistocene: do they distinguish types of interglacials? *Philosophical Transactions of the Royal Society, London*, B318: 539-557.

CATT, J.A. (1989). Relict properties in soils of the central and north-west European temperate region. *Catena Supplement*: 16, 41-58.

CATT, J.A. (Editor) (1990). Palaeopedology manual. *Quaternary International*, 6: 1-95.

CATT, J.A. (1991). Soils as indicators of Quaternary climatic change in mid-latitude regions. *Geoderma*, 51: 167-187.

CERLING, T.E. (1984). The stable isotopic composition of soil carbonate and its relationship to climate. *Earth and Planetary Science Letters*, 71: 229-240.

CERLING, T.E. & HAY, R.L. (1986). An isotopic study of palaeosol carbonates from Olduvai Gorge. *Quaternary Research*, 25: 63-78.

CHADWICK, O.A., BRIMHALL, G.H. & HENDRICKS, D.M. (1990). From a black to gray box - a mass balance interpretation of pedogenesis. *Geomorphology*, 3: 369-390.

CHESWORTH, W. (1973). The residua system of chemical weathering: a model for the chemical breakdown of silicate rocks at the surface of the earth. *Journal of Soil Science*, 24: 69-81.

DAN, J. & YAALON, D.H. (1982). Automorphic soils in Israel. In: D.H. Yaalon (Editor), *Aridic soils and geomorphic processes. Catena Supplement* 1: 103-115.

DUCHAUFOUR, P. (1982). *Pedology, pedogenesis and classification*. G. Allen and Unwin, London, 448 pp.

ERHART, H. (1932). Sur la nature et la genèse des paléo-sols du loess ancien d'Alsace. *Comptes Rendus de l'Academie des Sciences,* Paris, 194: 554-556.

FAO-UNESCO (1974). Soil map of the world 1:5 000 000: Volume 1, Legend. UNESCO, Paris, 59 pp.

FAO-UNESCO (1988). Soil map of the world revised legend. FAO, Rome, 119 pp.

HARDEN, J.W. (1982). A quantitative index of soil development from field descriptions: examples from a chronosequence in central California. *Geoderma*, 28: 1-28.

HODGSON, J.M. (1976). Soil Survey field handbook. *Soil Survey Technical Monograph* 5, Rothamsted Experimental Station, Harpenden, 99 pp.

HOOSBEEK, M.R. & BRYANT, R.B. (1992). Towards the quantitative modelling of pedogenesis - a review. *Geoderma*, 55:183-210.

HUGGETT, R.J. (1975). Soil landscape studies: a model of soil genesis. *Geoderma*, 13: 1-22.

HUIJZER, A.S. (1993). Cryogenic microfabrics and macrostructures. Interrelations, processes and palaeoenvironmental significance. Academisch Proefschrift Vrije Universiteit Amsterdam, 245 pp.

JENNY, H. (1941). *Factors of soil formation*. McGraw-Hill, New York, 281 pp.

JENNY, H. (1980). *The soil resource, origin and behavior*. Springer Verlag, New York, 377 pp.

JENNY, H. & LEONARD, C.D. (1935). Functional relationships between soil properties and rainfall. *Soil Science*, 38: 363-381.

JOHNSON, D.L. & HOLE, F.D. (1994). Soil formation theory: a summary of its principal impacts on geography, geomorphology, soil-geomorphology, Quaternary geology and paleopedology. In: R. Amundson, J. Harden and M. Singer (Editors), *Factors of soil formation: a fiftieth anniversary retrospective*. Soil Science Society of America Publication, 33: pp. 111-126.

JOHNSON, D.L., KELLER, E.A. & ROCKWELL, T.K. (1990). Dynamic pedogenesis: new views on some key soil concepts, and a model for interpreting Quaternary soils. *Quaternary Research*, 33: 306-319.

KEMP, R.A. (1987). Genesis and environmental significance of a buried Middle Pleistocene soil in eastern England. *Geoderma*, 41: 49-77.

KUBIENA, W.L. (1953). The soils of Europe: illustrated diagnosis and systematics. T. Murby, London, 318 pp.

LARSEN, G. & CHILANGAR, G.V. (Editors) (1967). Diagenesis in sediments. Developments in Sedimentology 8. Elsevier, Amsterdam, 551 pp.

LEVINE, E.R. & CIOLKOSZ, E.J. (1986). A computer simulation model for soil genesis applications. *Soil Science Society of America Journal*, 50: 661-667.

MARION, G.M., SCHLESINGER, W.H. & FONTEYN, P.J. (1985). CALDEP: a regional model for soil formation in south western deserts. *Soil Science*, 139: 468-481.

McKEAGUE, J.A. (1983). Clay skins and argillic horizons. In: P. Bullock and C.P. Murphy (Editors), *Soil Micromorphology* Volume 1. AB Academic Publishers, Berkhamsted: pp. 367-387.

McKEAGUE, J.A., MacDOUGALL, J.I. & MILES, N.M. (1973). Micromorphological, physical, chemical and mineralogical properties of a catena of soils from Prince Edward Island in relation to their classification and genesis. *Canadian Journal of Soil Science*, 53: 281-295.

McSWEENEY, K., SLATER, B.K., HAMMER, R.D., BELL, J.C., GESSLER, P.E. & PETERSON, G.W. (1994). Towards a new framework for modelling the soil-landscape continuum. In: R. Arundson, J. Harden and M. Singer (Editors), *Factors of soil formation: a fiftieth anniversary retrospective*. Soil Science Society of America Special Publication 33, pp. 127-145.

MÜCHER, H.J. & MOROZOVA, T.D. (1983). The application of soil micromorphology in Quaternary geology and geomorphology. In: P. Bullock and C.P. Murphy (Editors), *Soil Micromorphology* Volume 1. AB Academic Publishers, Berkhamsted: pp. 151-194.

MÜCKENHAUSEN, E. (1982). Die Bodenkunde und ihre geologischen, geomorphologischen, mineralogischen und petrologischen Grundlagen. D.L.G. Verlag, Frankfurt, 632pp.

NIKIFOROFF, C.C. (1943). Introduction to paleopedoloy. *American Journal of Science*, 241: 194-200.

POLYNOV, V.V. (1927). Contributions of Russian scientists to palaeopedology. In: *Russian pedological investigations*, Part 8. Academy of Science, Moscow.

RAMANN, E. (1911). *Bodenkunde* (3rd Edition). J. Springer, Berlin, 431 pp.

RAMANN, E. (1928). *The evolution and classification of soils* (Trans. C.L. Whittles). W. Heffer and Sons Ltd., Cambridge, 127 pp.

RETALLACK, G.J. (1990). Soils of the past: an introduction to palaeopedology. Unwin and Hyman, New York, 520 pp.

RETALLACK, G.J. (1994). The environmental factor approach to the interpretation of paleosols. In: R. Amundson, J. Harden and M. Singer (Editors), *Factors of soil formation: a fiftieth anniversary retrospective*. Soil Science Society of America Publication 33, pp. 31-64.

RUHE, R.V. (1956). Geomorphic surfaces and the nature of soils. *Soil Science*, 82: 441-455.

RUHE, R.V. (1984). Soil climate system across the prairies in the midwestern USA. *Geoderma*, 34: 204-219.

RUHE, R.V. & OLSON, C.G. (1980). Soil welding. *Soil Science*, 130: 132-139.

RUNGE, E.C.A. (1973). Soil development sequence and energy models. *Soil Science*, 115:183-193.

SMECK, N.E., RUNGE, E.C.A. & MACKINTOSH, E.E. (1983). Dynamics and genetic modelling of soil systems. In: L.P. Wilding (Editor), *Pedogenesis and soil taxonomy I. Concepts and interactions*. Developments in Soil Science IIA, Elsevier, Amsterdam, pp. 51-81.

SOIL SURVEY STAFF (1951). Soil survey manual. USDA Agriculture Handbook 18. US Government Printing Office, Washington, D.C., 503 pp.

SOIL SURVEY STAFF (1975). Soil Taxonomy. A basic system of soil classification for making and interpreting soil surveys. USDA Agriculture Handbook 436. US Government Printing Office, Washington, D.C., 754 pp.

SOIL SURVEY STAFF (1992). Keys to soil taxonomy. Soil Management Support Services Technical Monograph 19 (5th Edition). Pocahontas Press, Blacksburg, Virginia, 541 pp.

TANDARICH, J.P. & SPRECHER, S.W. (1994). The intellectual background for the factors of soil formation. In: R. Amundson, J. Harden and M. Singer (Editors), *Factors of soil formation: a fiftieth anniversary retrospective*. Soil Science Society of America Publication 33:1-13.

VALENTINE, K.W.G. & DALLYMPLE, J.B. (1975). The identification, lateral variation, and chronology of two buried palaeocatenas at Woodhall Spa and West Runton, England. *Quaternary Research*, 5: 551-590.

VALENTINE, K.W.G. & DALRYMPLE, J.B. (1976). Quaternary buried palaeosols: a critical review. *Quaternary Research*, 6: 209-222.

VAN VLIET-LANOË, B. (1985). Frost effects in soils. In: J. Boardman (Editor), *Soils and Quaternary landscape evolution*. J. Wiley, Chichester: pp. 117-158.

VREEKEN, W.J. (1975). Principal kinds of chronosequences and their significance in soil history. *Journal of Soil Science*, 26: 378-394.

WALKER, T.R. (1967). Formation of red beds in modern and ancient deserts. *Bulletin of the Geological Society of America*, 78: 353-368.

WASHBURN, A.L. (1979). Geocryology. A survey of periglacial processes and environments. E. Arnold, London, 406 pp.

WHITESIDE, E.P. (1959). A proposed system of genetic soil horizon designations. *Soils and Fertilizers*, 22: 1-8.

WILDING, LP. & DREES, L.R. (1983). Spatial variability and pedology. In: L.P. Wilding (Editor), *Pedogenesis and soil taxonomy I. Concepts and interactions*. Developments in Soil Science IIA, Elsevier, Amsterdam, pp. 83-116.

YAALON, D.H. (1971). Soil-forming processes in time and space. In: D.H. Yaalon (Editor), *Palaeopedology: origin, nature and dating of palaeosols*. International Soil Science Society and Israel Universities Press, pp. 29-39.

YAALON, D.H. (1975). Conceptual models in pedogenesis: can soil-forming functions be solved? *Geoderma*, 14: 189-205.

Quaternary Proceedings No. 5, 1993 69-81
© Quaternary Research Association, Cambridge.

Loess-Palaeosol-Sequences in Tadjikistan as a Palaeoclimatic Record of the Quaternary in Central Asia.

A. Bronger, R. Winter, O. Derevjanko & S. Aldag

A. Bronger, R. Winter, O. Derevjanko & S. Aldag, 1995 Loess-palaeosol-sequences in Tadjikistan as a palaeoclimatic record of the Quaternary in Central Asia, In *Wind Blown Sediments in the Quaternary Record* (Edward Derbyshire). Quaternary Proceedings No. 4, John Wiley & Sons Ltd., Chichester, pp. 69-81.

Abstract

Loess profiles in Tadjikistan contain numerous palaeosols and have been described from a chronostratigraphical point of view; for example, the B/M boundary was recently confirmed between pedocomplexes PK IX and X. However, no studies regarding the genesis of the palaeosols, essential for palaeoclimatic deductions, have been made so far. A genetic classification is difficult because almost all fossil soils are truncated and recalcified from the overlying loess. Micromorphology allows primary and secondary carbonates to be distinguished and provides unequivocal evidence of the process of clay illuviation. The grain-size distribution and 160 thin sections of loesses and palaeosols were studied from the upper and central part of the Karamaydan and the central and lower part of the Chashmanigar sequences. Typical loess occurs in all parts of the Karamaydan and Chashmanigar profiles down to PK XXVII. All strongly developed B or Bt horizons in both exposures represent interglacials similar to the Holocene. The loess-palaeosol-sequence at Karamaydan (PK I-X) can very well be compared with the $\delta^{18}O$ - record shown in the SPECMAP curve of the Brunhes Chron. From the present knowledge the loess-palaeosol-sequence at Chashmanigar (PK X - PK XXX) provides the most detailed sequence for the Matuyama Chron in Central Asia. It gives even more palaeoclimatic information than the loess profiles in China and even the deep-sea curves known so far.

KEYWORDS: pedocomplexes; micromorphology (soils); grain-size distributions; SPECMAP.

A. Bronger, R. Winter and S. Aldag, Department of Geography, University of Kiel,
D-24098 Kiel, Germany.

O. Derevjanko, Tadjik Geological Survey, Krasnych Partizan, 27, 734025 Dushanbe, Tadjikistan.

Introduction and Approach

Loess profiles in the former Soviet Central Asia, especially in the Tadjik Depression, contain numerous palaeosols. They are up to 90 m thick in the Tashkent region and up to 150 - 200 m in the Tadjik Depression. They have been described from a geological-stratigraphical point of view, mainly by Dodonov (e.g. 1979, 1984, 1991), Dodonov *et al.* (1977, 1982), Lazarenko (1977, 1984) and Lazarenko *et al.* (1981). Penkov and Gamov, (1977, 1988) identified the Brunhes/Matuyama (B/M) boundary in several loess profiles mainly in Tadjikistan between the pedocomplexes IX and X (Fig. 4, see below). The Jaramillo- and Olduvai subchrons could also be found in some profiles (Penkov & Gamov 1977, 1980; Dodonov & Penkov 1977; Lazarenko *et al.* 1991). The palaeomagnetism of the loess profile at Karamaydan (Fig. 3, 4) was recently investigated in greater detail by Förster and Heller (1994, in press). They confirmed the stratigraphic position of the B/M-boundary between PK IX and PK X.

However, very little attention has been paid to the *genesis* of the palaeosols, which is necessary for palaeoclimatic deductions. Dodonov (1991, Figs. 4, 5) differentiated "buried soil", "embryonic soil" and "pedocomplex", and Lazarenko (1984, Fig. 13-1) differentiated "fully developed zonal-type

buried soils", "moderately developed buried soils" and "slightly developed and embryonic soils and transitional layers of soil-sediment type". But they did not consider genesis and classification of the palaeosols. Sometimes different soil horizons were identified in palaeosols or pedocomplexes (Lazarenko 1984, Fig. 13-2; Lomov & Ranov 1985), but without further description or evidence of the process of clay illuviation by micromorphology they do not allow inferences about their origin and climatic significance. No Ck horizons were mentioned by these authors and shown in the above mentioned schemes although distinct calcareous nodules ("Lößkindl") are a very characteristic feature of many soils (cf. Figs. 3-5). They are especially common in the loess profile of Chashmanigar, which is considered "the Eopleistocene (= Old Pleistocene) stratotype of Southern Tadjikistan loess formation" by Dodonov and Lomov (1985, 223). Lazarenko (1984, Fig. 13-2) mentioned a "dense epigenetic lime crust" a few centimetres thick in the uppermost part of a BCca horizon of his "fully developed buried soils". - From a few samples of loesses and paleosols grain-size analyses were done (Dodonov 1991, Fig. 6). - The palaeosols were not compared with the Holocene "zonal" or climaphytomorphic soil (Schroeder, 1984) of this loess area. These soils at an altitude of about 1000-1800 m a.s.l. were earlier called "Cinnamonic Brown

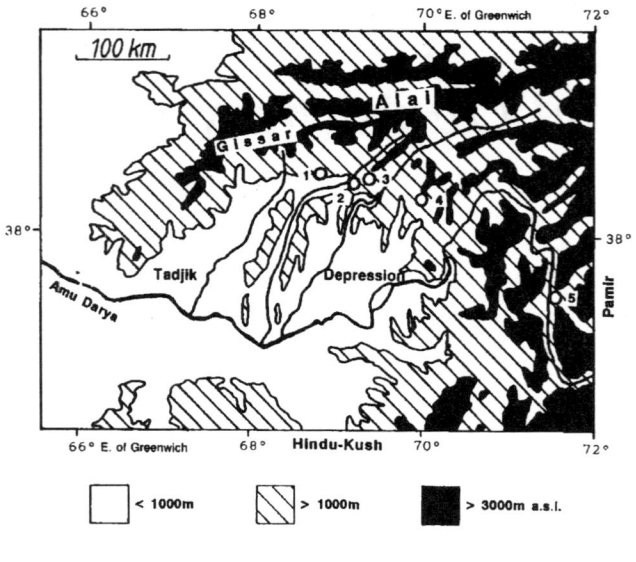

Figure 1. Location map of the Tadjik Depression.

soils" (Lomov & Sosnovskaya 1977). Later they were divided into "Calcareous", "Typic" and "Leached Cinnamonic soils" (Lomov 1985). Even the last was regarded as a "subtropical steppe soil". For comparison with the buried palaeosols, especially to draw palaeoclimatic conclusions, it is important to study the genesis of the Holocene soil of the area in relation

to the soil forming factors. Consequently we briefly consider the two most important soil forming factors, climate and vegetation, in SW Tadjikistan.

The *climate* is characterized by cool winters and very warm summers. The precipitation is distinctly seasonal, with moist winters and very dry summers (Fig. 2). Superimposition of the temperature and the precipitation regimes results in a xeric soil moisture regime (Soil Survey Staff, 1975: 51-57; Van Wambeke *et al.* 1986). A considerable water deficit in the (late) summer is mitigated by a water utilization because of the high plant available water capacity of loess soils. In the late autumn and winter a water recharge occurs to give a large water surplus (Fig. 2). The meterological station Chormazak Pass (Figs. 1,2) is near to Karatau where we investigated two profiles of the Holocene soil (Fig. 3, the two upper soils). Fayzabad is close to the loess profile of Karamaydan, but is 400 - 500 m less in elevation. Khovaling is situated about 15 - 20 km ESE of the loess profile of Chashmanigar, and is about 1650 - 1750 m a.s.l. Towards the southeast near the Pamir the precipitation decreases markedly as shown for Chorog (Figs. 1 ,2).

In the last 100 years the landscape has been strongly degraded and severe soil erosion has taken place, because of a large increase in the population. For example the central parts of both exposures we investigated, Karamaydan and Chashmanigar, were covered by big landslides in spring 1992. Because of the anthropogenic degradation of the vegetation the former existence of forests between 1000 - 2000 m a.s.l. has been doubted (Lomov 1985). However, Zapriagaeva (1964, 1968) and Staninkovitch (1968) concluded that the potential

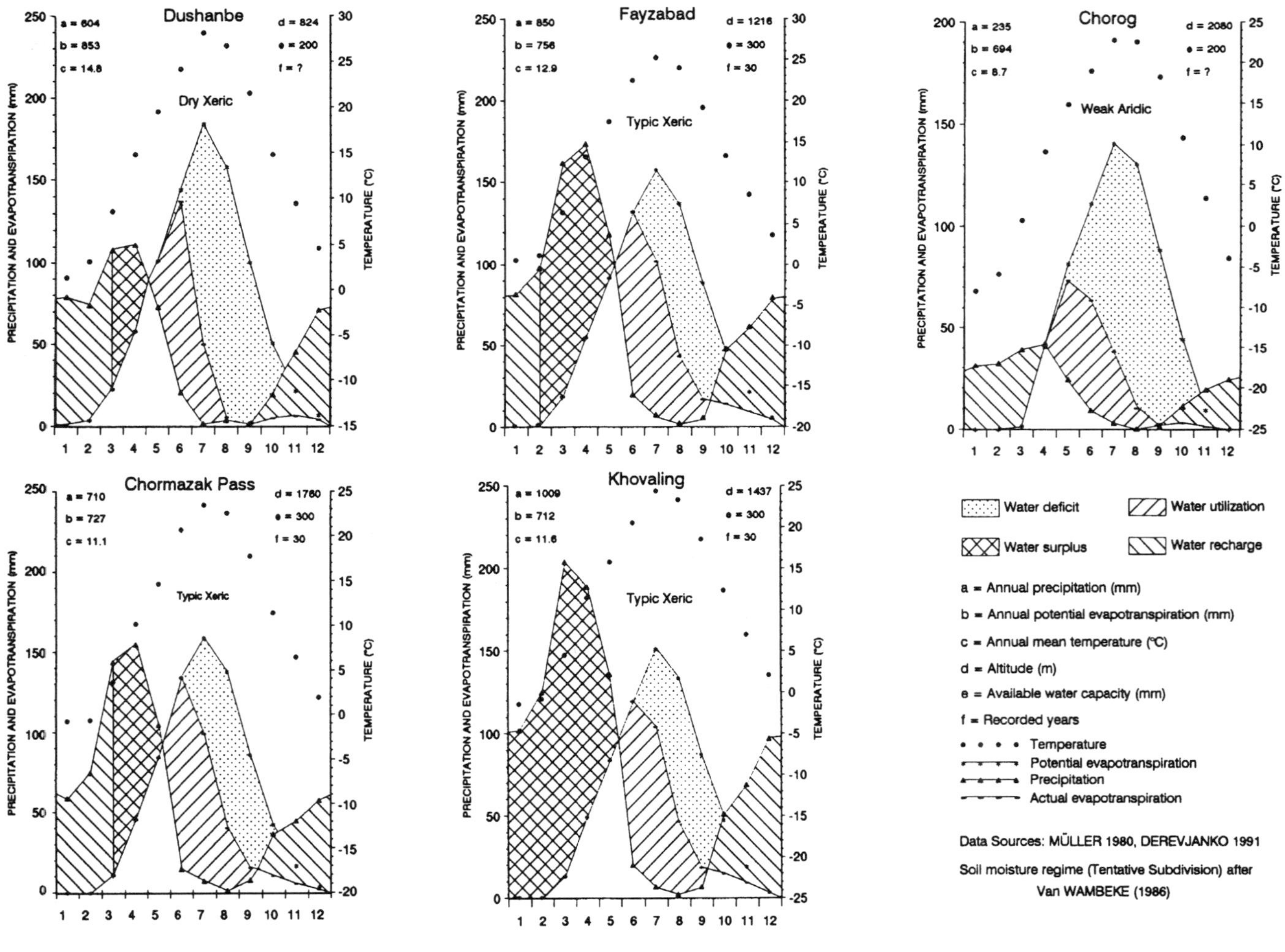

Figure 2. Climatic data and soil water balance of selected stations in the loess area of SW Tadjikistan.

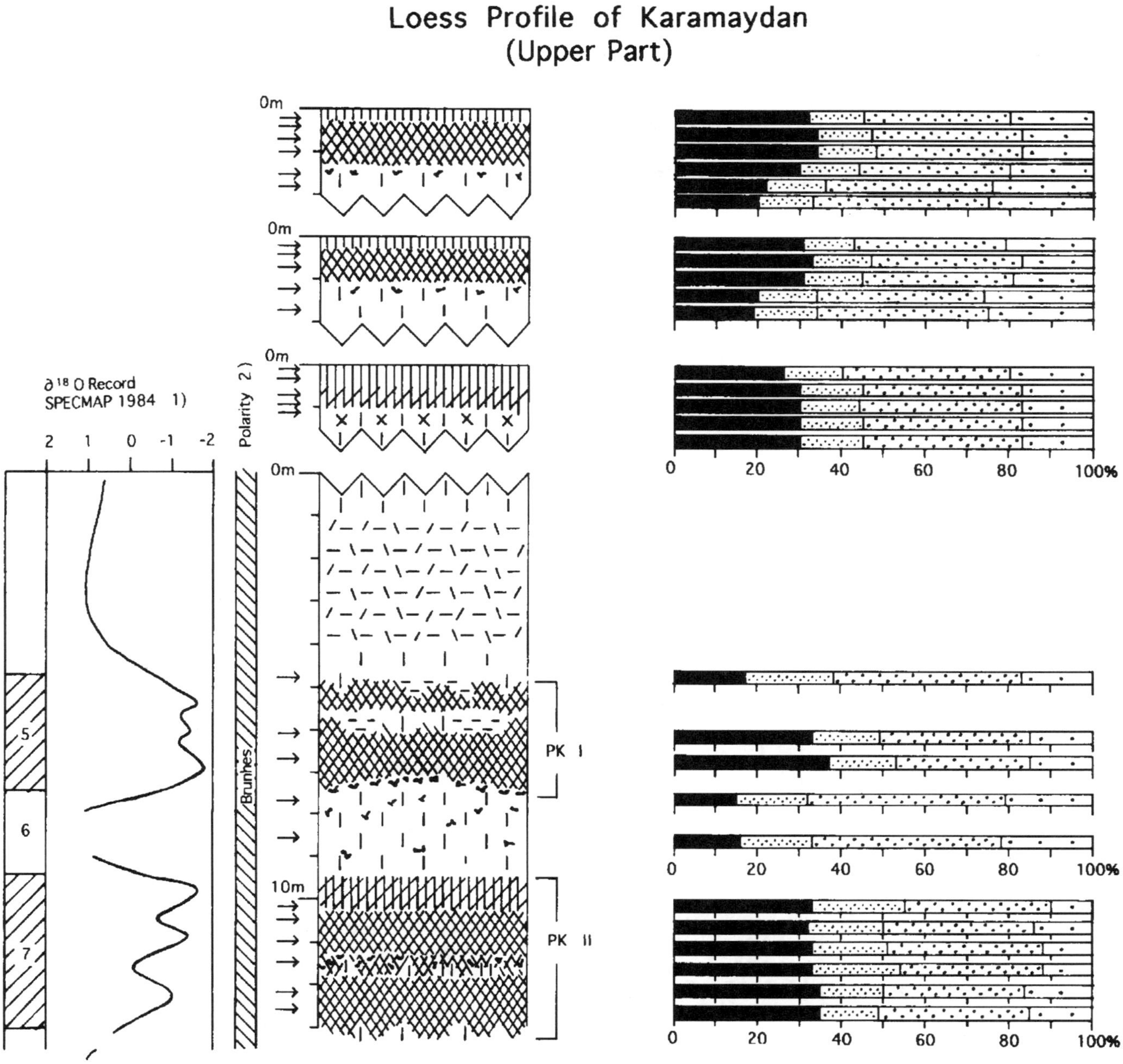

Loess Profile of Karamaydan
(Upper Part)

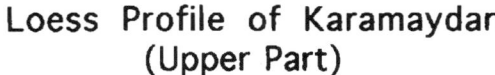

Figure 3. Genesis of the soil horizons of three selected Holocene soils and upper Pleistocene pedocomplexes PK I and II and particle size distribution of the loess profile of Karamaydan, Tadjikistan, eastern section and a comparison with the adjusted SPECMAP curve.

natural vegetation in Tadjikistan south of the Tien-shan and west of the Pamir in 1000 - 2200 (2400) m a.s.l. was a *specific broad-leaved forest* consisting of maples (*Acer turkestanicum*), *Platanus orientalis, Fraxinus potamolina*, walnut (*Juglans regia*) and many other wildgrowing fruit trees such as almond (*Amygdalus communis, A. bucharica*) and different species of *Celtis, Pyrus, Crataegus* etc. In several parts also junipers (*Juniperus seravschanica*) occur at elevations of 1700 - 2300 m.

Materials and Methodological Aspects

In several summers in the second half of the 1980s field surveys were made near the Karamaydan, Chashmanigar and Khonako (near Khovaling) sections by O. Derevjanko. – To draw palaeoclimatic deductions from loess-palaeosol-sequences we selected several sections of the loess profile at *Karamaydan*. This profile seems to show the most complete loess-palaeosol-sequence of the Brunhes Chron in Tadjikistan. An angular

unconformity below PK X in Karamaydan as well as in Chashmanigar confirms the stratigraphic correlation between both profiles. We also focussed attention on the central and lower parts of one of the sections at *Chashmanigar* which we now know shows the most detailed loess-palaeosol-sequence of the Matuyama chron in Central Asia. It is even more detailed than that of the Luochuan loess profile (Bronger & Heinkele 1989a, Fig. 2) or in Xifeng (Guo *et al.* 1991), both in the Central Loess Plateau in China.

For loess-palaeosol stratigraphy and palaeoclimatic deductions it is necessary to classify buried palaeosols (Catt 1994) genetically and typologically, a procedure which allows comparisons between profiles. Observations in the field are not always sufficient, partly because most of the buried soils are truncated. Some important diagnostic features which can be recognized in modern soils, *e.g.* chemical properties, have also disappeared or have been changed because of post-pedogenetic alteration of the palaeosols (Catt 1988). For

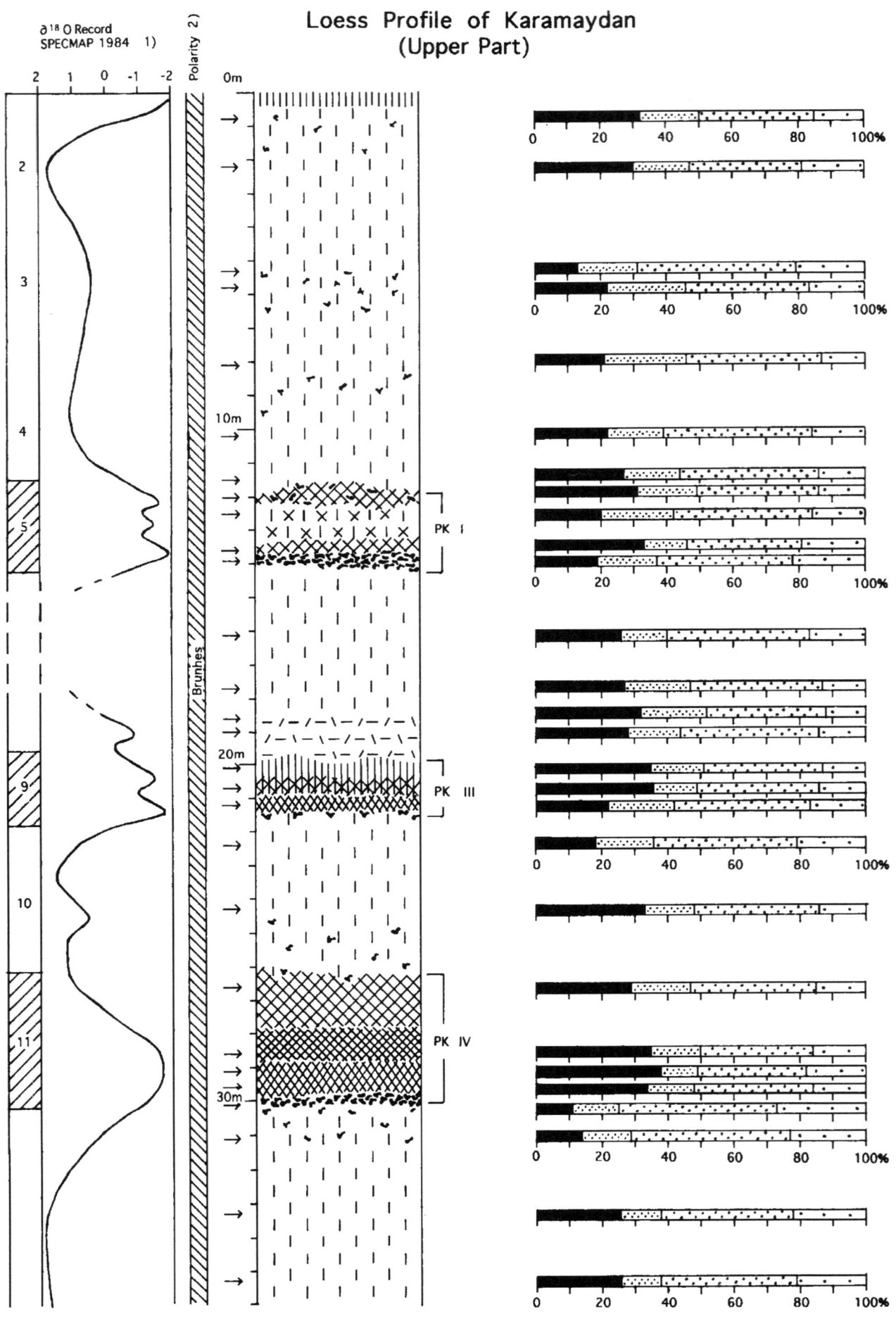

Loess Profile of Karamaydan
(Upper Part)

Figure 4. (caption p. 74)

Draft: R. Winter, O. Derevjanko, A. Bronger

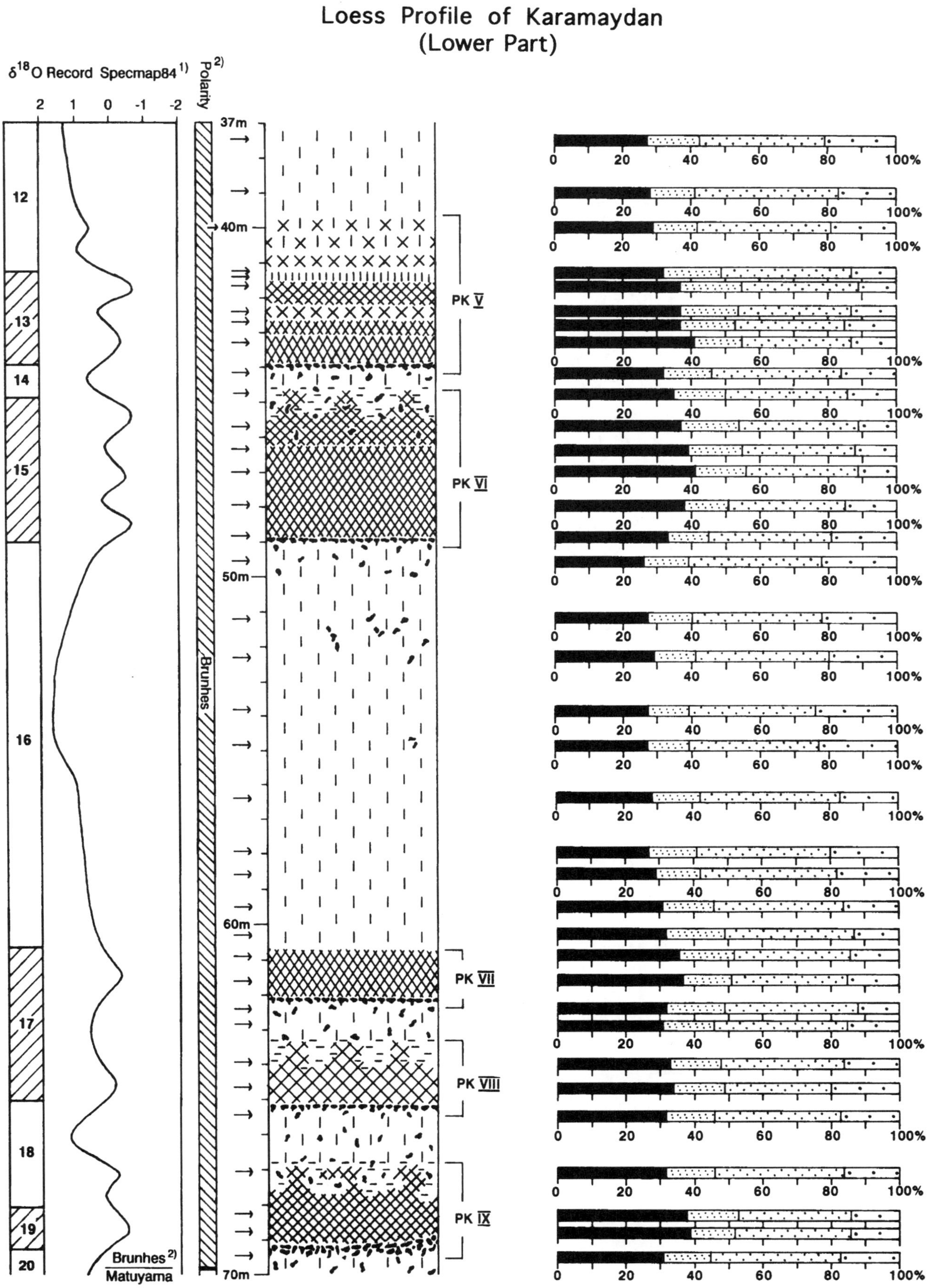
Loess Profile of Karamaydan
(Lower Part)
δ18O Record Specmap84 1)
Polarity 2)
2 1 0 -1 -2
12
13
14
15
16
17
18
19
20
Brunhes 2)
Matuyama
Brunhes
37m
40m
50m
60m
70m
PK V
PK VI
PK VII
PK VIII
PK IX
0 20 40 60 80 100%

Figure 4. (caption p. 74)

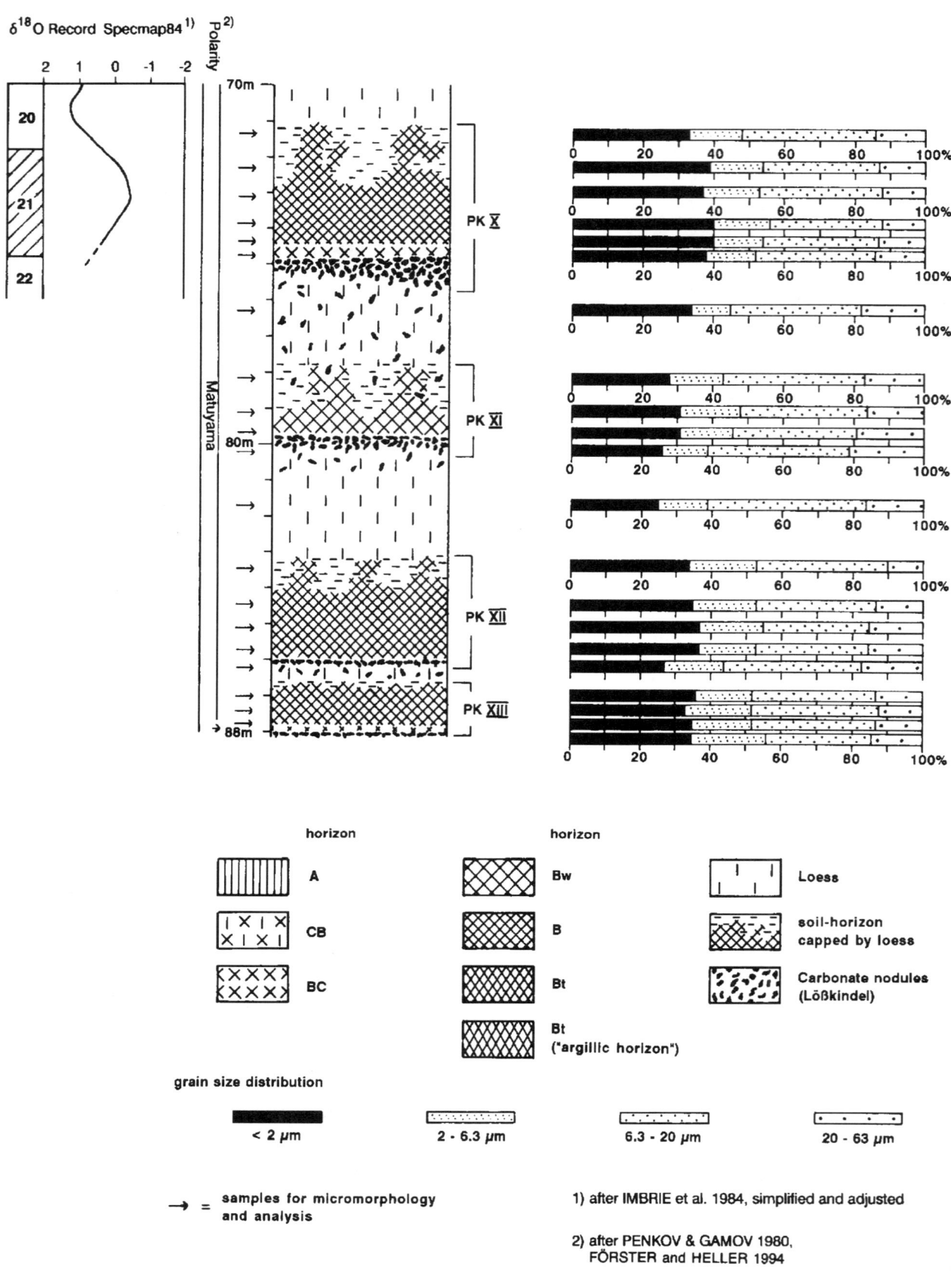

Figure 4. Genesis of the soil horizons of pedocomplexes PK I and III - XIII and particle size distribution of the loess profile of Karamaydan, western section and a comparison with the adjusted SPECMAP curve.

Loess Profile of Chashmanigar
(Central Part)

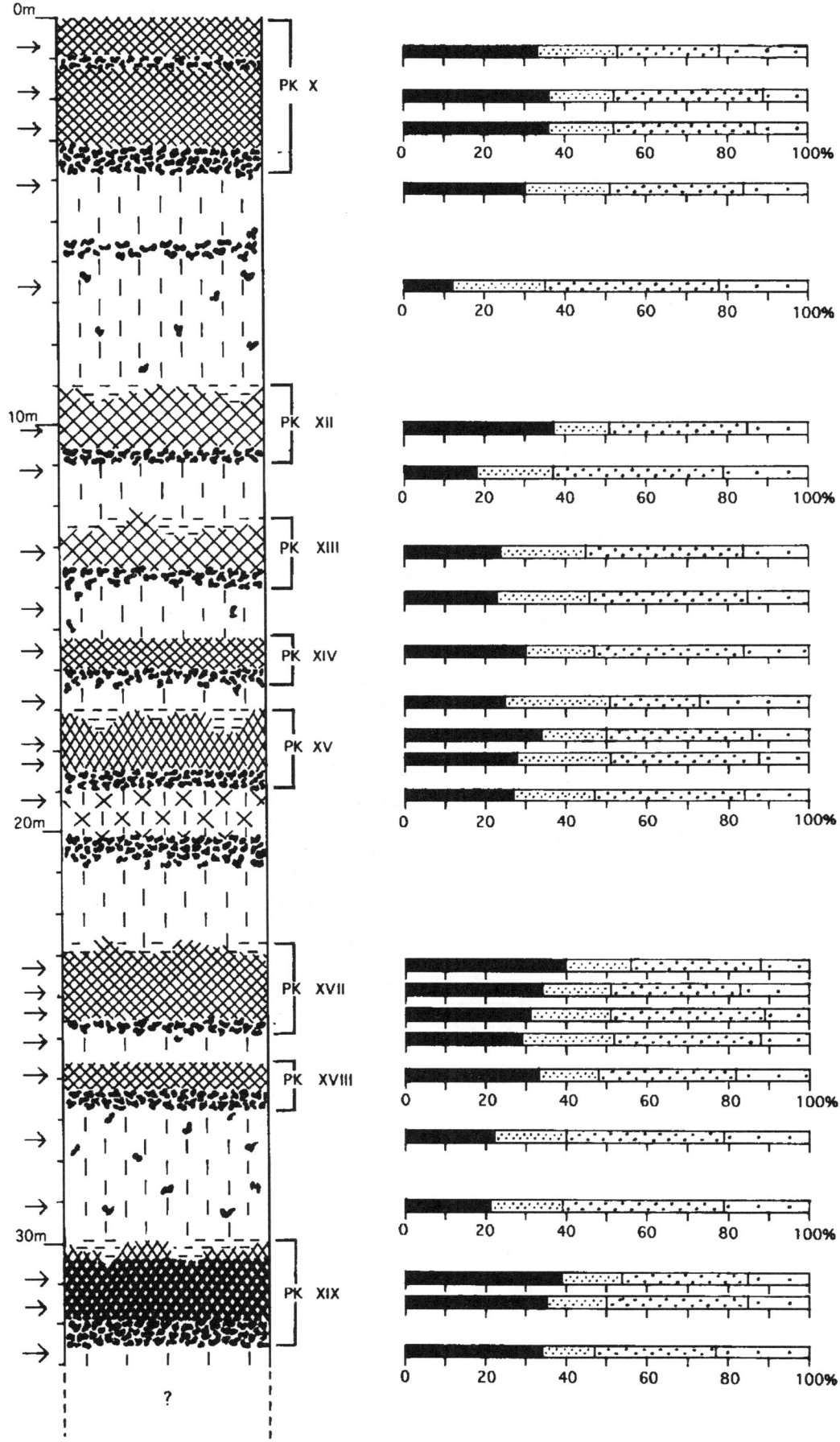

Figure 5. (caption p. 76)

Loess Profile of Chashmanigar
(Lower Part)

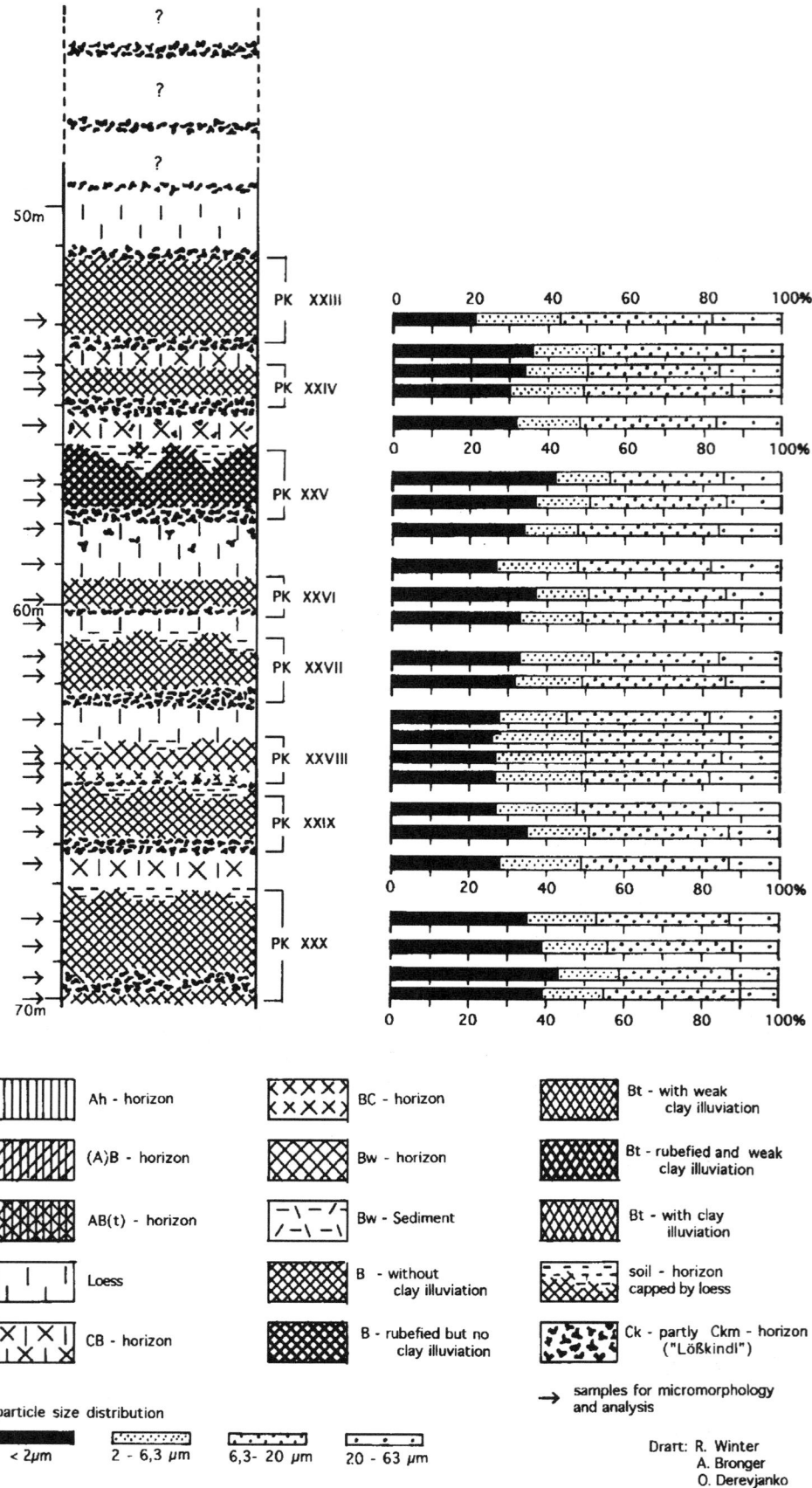

Figure 5. Genesis of the soil horizons of pedocomplexes PK X, XII-XIX and XXIII - XXX and particle size distribution of the loess profile of Chashmanigar, Tadjikistan, central and lower part.

instance progressive decomposition of organic material with advancing age during the upper Pleistocene (Bronger 1966, 1974) and even early Holocene (Bronger & Heinkele 1989a) prevents us designating most buried Chernozems and Phaeozems as Mollisols in the Soil Taxonomy (Soil Survey Staff, 1975). Another example in loess areas is that all buried soils are more or less *recalcified* from the overlying loess, so in contrast to modern soils pH- or base saturation values, although quoted even recently (Guo *et al.* 1991), are of no use. It is therefore necessary to study the sequence of *processes* in buried soils to provide information about pedogenesis as well as diagenesis. *Micromorphology* is of special significance, for it provides a complete view of soil development through different stages even in polygenetic pedocomplexes (Kubiena 1959, 1964; Smolikova 1967, 1971, 1990; Bronger 1969/70, 1976; Catt 1990) and often reveals the effects of diagenetic processes after burial. Micromorphology can, for example, distinguish between secondary and primary carbonates. *Secondary carbonates* occur in voids as fine calcite concentrations of micrite and microsparite (Courty *et al.* 1987), as accumulations of coarse calcite in root channels, or as calcite needles in finer pores (*e.g.* Bronger 1966: photos 9-12, 1976: photos 10, 18, 19; Bronger *et al.* 1987: figs. 2-4, 1994: figs. la-c; Bronger & Heinkele 1989a: fig. 9; Wieder & Yaalon 1982); macroscopically these forms are called pseudomycelium. However, individual carbonate clasts especially of coarse silt-size, irregularly distributed in the soil matrix, can be regarded as *primary* (Bronger 1976: photos 9, 17 and 23). This genetic distinction between carbonate constituents is of special importance; for example it allows us to distinguish between Degraded Chernozems now Phaeozems (Bronger 1994) and Chernozems still containing primary carbonates, the latter developed in a shorter time and a drier climate.

Micromorphology also provides an unequivocal indication of *clay illuviation*. This process takes place in the pH range 4.5-6 (0.1 M KCl or CaCl$_2$) in soils on plateau sites well above the groundwater table. It is associated in temperate climates with a forest vegetation and udic soil moisture regime, but not with a steppe vegetation and ustic soil moisture regime (for definitions see Soil Survey Staff, 1975: 51-57)if Na$^+$-ions are absent. This relationship has been found in many Holocene loess soils in east Central Europe, in the western part of the Central Lowlands and in the Great Plains of the USA (Bronger 1976: 35-106; 1991). Even when the illuviation argillans (Brewer 1964; Bullock *et al.* 1985) as unequivocal signs of clay illuviation are transformed by changing environmental conditions in buried or relict soils (surface paleosols, Catt 1994), they still remain visible as a very stable element of the fabric. They are even found in the excrements of earthworms (*e.g.* Zachariae 1964; Bronger 1976: photo 3; Catt 1989: photo IA) or in "Fließerde" (gelifucted soil) material from a buried Mid-Pleistocene Udalf (Bronger 1969/70: fig. 2A, B; Catt 1989: photo IB-D). Other features of illuvial clay may be less clear; interpretations are subjective (McKeague 1983; Bullock & Thompson 1985; Bronger 1991: 44). However, it is necessary to separate the basically climate-controlled process of clay illuviation, in the form of illuviation argillans, from the process of argillipedoturbation, in the form of oriented birefringence (*e.g.* Bronger *et al.* 1994, fig. 6d), which is predominantly a lithogenic process determined by the soil material especially in vertic soils *sensu lato*.

From the eastern (Fig. 3) and western (Fig. 4) sections of Karamaydan, from the central and lower part of Chashmanigar (Fig. 5) and from three selected Holocene soils, 160 oriented

samples were taken for micromorphological investigations, also 160 bulk samples for soil-physical and -chemical analyses. Sampling points are shown in Figs. 3-5. In this first approach results and interpretations were based upon field characteristics, micromorphology and grain-size distribution, the latter also indicating the amount of pedochemical clay mineral formation. The results of the mineralogical investigations will be given in a latter paper. Grain-size distribution was determined by the pipette method after destruction of organic material by 6% H$_2$O$_2$ and of CaCO$_3$ by 0,1 M HCl; particles > 63µm were absent or <1%. The genetic designations of the soil horizons follow the international convention (FAO – Unesco 1974; Wilding *et al.* 1983) as much as possible. Weakly developed horizons or subhorizons are put in parenthesis.

Results and Interpretation

Results of the grain-size distribution, field studies and micromorphological investigations and their interpretation for the sections at Karamaydan and Chashmanigar and for three Holocene soils are shown in Figs. 3-5. A few points are emphasized briefly.

Two of the *Holocene soils*, previously called "Leached Cinnamonic Soils", and sited near Karatau at about 1700 m a.s.l. in a plateau position not far from the Chormazak Pass (Figs. 1,2), are represented in Fig. 3 (uppermost two profiles). The two soils are free of calcium carbonate, and the pH-values (measured in 1 M KCl) were as low as 4.9. Both have a dark grey Ah horizon, 27 cm in thickness which meets the requirement of a mollic epipedon or mollic horizon (Soil Survey Staff, 1975; ISSS-ISRIC-FAO, 1994). Their Bt horizons - with 7.5 YR hues in the Munsell notation - are 105 cm and 80 cm thick, respectively. Both have a sharp boundary to a whitish Ck horizon, which is typical of forest soils, and not of steppe soils. Micromorphologically several parts of the Ah horizons have a fine spongy fabric rich in aggregates and pores, indicating a high biological activity. This partly could be because forest has been replaced by steppe because of human activity. In the Bt horizons this type of fabric can only be seen in some parts, most parts have a more dense fabric with much less aggregates and voids. Illuviation argillans are only visible locally and form much less than 1% by area. Also the very small increase in clay content (Fig. 3) in both soils is not sufficient to qualify these subsurface horizons as argillic horizons in *Soil Taxonomy* (Soil Survey Staff, 1975) or as argic horizons in the *World Reference Base* (ISSS-ISRIC-FAO, 1994). These broad-leaved forest soils (cf. Introduction) have to be classified as *Typic Haploxerolls* or *Haplic Phaeozems* (Bronger 1994). It is surprising that the process of clay illuviation has scarely occurred despite the considerable water surplus in March and April (Fig. 2). However, we found similar relationships in the Kashmir Valley (India) also in a xeric soil moisture regime: the soil water balance of Srinagar (Bronger *et al.* 1987: Fig. 6) is similar to that of the Chormazak Pass and the process of clay illuviation is also very weak in the Holocene loess soils of this area. More research on Holocene soils, especially in the more moist Khovaling - Chashmanigar area is needed. The Holocene soil at Karamaydan (Fig. 3, third soil from the top) is laterally recalcified because of its situation on a flat slope.

In the loess sections of Karamaydan and Chashmanigar the palaeosols are grouped into *pedocomplexes* (PK's) as done by Dodonov and Lazarenko. This concept originates from Czechoslovakia (*e.g.* Smolikova 1967). Pedocomplexes are recently redefined by the Palaeopedology Commission of INQUA. They "comprise two or more soils (pedomembers), which are separated over large areas by thin (almost) unmodified deposits and are overlain and underlain by greater thickness of unmodified deposits or by unconformities" (Catt 1994). Although truncated, most of the pedocomplexes contain two or more soils or soil horizons, each transformed by pedogenic processes from newly deposited loess or loess-like deposits. The designations of our pedocomplexes differ partly from those of Dodonov (1991) and Lazarenko (1984; see Conclusions and Discussion).

Typical loess, which is dominated micromorphologically by a massive structure, occurs in all parts in the Karamaydan profile and in the central and lower part of the profile of Chashmanigar to below our PK XXVII (see below). All loesses contain plenty of *primary* and *secondary* calcite in different amounts. The latter are a sign of high bicarbonate metabolism as a result of a less arid climate. Several loesses in the middle and upper part of the profile of Karamaydan have very large amounts of primary calcite and much less secondary calcite because of more arid conditions. In addition to various amounts of clay (< 2 μm), between 6 and 34 % (Figs. 3-5), all loesses consist of silt with little or no sand. In the silt fraction *medium silt* (6-20 μm) is dominant in every sample, ranging from 34 to 48 % of the total, nearly 40% on average even in the lower part of the Chashmanigar profile. This contrasts with the loesses of central Europe where coarse silt (20-63 μm) is dominant (*e.g.* Bronger 1966: Figs. 2, 3; 1976: Tables 7-10). But the loesses of the Kashmir Valley / India also show a dominance of medium silt (Bronger *et al.* 1986). The high content of medium silt leads to the conclusion that loess in the Tadjik Depression was partly derived from more distant areas.

In the *pedocomplexes* one can find almost all transitions between loess and a strongly developed rubefied Bt horizon (Figs. 3-5). For the characterization of the soil horizons the differentiation of primary and secondary calcite was especially important, because *all* soils were secondarily enriched with lime. CB horizons in which the original properties of loess dominate those of a B horizon, still contains primary calcite, whereas BC horizons do not contain primary calcite. A CB horizon is normally B horizon material mixed with loess for example by solifluction. For instance below the distinct Ck or Ckm horizon of PK XV in the central part of the Chashmanigar exposure and an even more distinct Ckm (nodule) horizon below (Fig. 5) only a CB horizon occurs, probably as a remnant of a well developed soil (because of the strongly developed Ckm horizon). Therefore we designate the soil below – between 23 and 24,5 m – as PK XVII because largest parts of a PK XVI have been eroded. Between the distinct Ckm (nodule) horizon of PK X and another well expressed one almost 2 m below only typical loess occurs. Probably the soil belonging to the lower Ck or Ckm horizon (PK XI, Fig. 5) is totally eroded; therefore we designate the pedocomplex underneath – between 9 and 10,5 m – as PK XII. Above the distinct nodule horizons on top of PK XXIII (Fig. 5) there is probably a similar example: probably a PK XXII has been eroded (see below). The pedocomplexes XII and XIII at Chashmanigar consist only of rather weakly developed Bw horizons, although they are already leached (*i.e.* free of primary carbonates), however, also with well developed Ckm (nodule) horizons (Fig. 5). The soils (pedocomplexes) in the same stratigraphic position at Karamaydan (PK XII and XIII, see Fig. 4) show a well developed B horizon, though without any clay illuviation. Therefore both soils at Chashmanigar are probably only remnants of former more strongly developed soils. The much greater clay content (pedogenetically formed?) of PK XII (Fig. 5) compared with the loess underneath may be an additional indicator of this conclusion.

A similar example is provided by PK I at Karamaydan: it is well developed in the western part (Fig. 3) but only remnants are left in the eastern part (Fig. 4). However, several other pedocomplexes in both exposures also show *lithological discontinuities* which can only be indicated in this paper. Between 33 and 49 m depth at Chashmanigar the steep slope is covered by vegetation. However, two more distinct nodule ("Lößkindl") horizons can be seen, probably indicating two more palaeosols or pedocomplexes (PK XX and XXI, Fig. 5). Thick Ckm horizons contrast with loesses and soils in being very resistant to soil erosion, and so stand out as harder layers.

Besides the CB, BC, and Bw horizons we also distinguish different well developed B and Bt horizons, mainly according to the presence or absence of illuviation argillans. B horizons show no signs of clay illuviation. Bt horizons "with weak clay illuviation" do not show enough illuviation argillans for them to qualify as argillic or argic horizons in Soil Taxonomy or the World Reference Base (like the Bt horizons in the two modern soils). Only a few of the Bt horizons show (nearly) enough illuviation argillans to qualify as argillic or argic horizons: at Karamaydan the Bt horizons of the PK I and PK II, the lower parts of the PK III and PK V; at Chashmanigar only PK XV. The illuviation argillans are mostly aged, with decreased birefringence, and are partly reworked into the fabric. All B, Bt and even the weakly developed Bw horizons are within the 7,5 YR range of the Munsell notation. Only the Bt horizon with weak clay illuviation of PK XIX and the B horizon of PK XXV at Chashmanigar (Fig. 5) are more rubefied: their Munsell notation is in the 5 YR range. However, the single criterion of rubefication or formation of hematite at the expense of goethite is often exaggerated as an indicator of a warmer climate. Hematite formation needs a distinct seasonal climate with dry periods, good drainage and little or no humus.

All strongly developed B or Bt horizons in the two exposures represent *interglacials* similar to the Holocene. They were formed under *broad-leaved forests*. Those with an argillic horizon developed in a slightly moister climate. Several have a thicker B or Bt horizon and a more strongly developed Ckm horizon, indicating that the interglacial may have lasted for a longer time than the Holocene. The thick but weakly developed Bw horizons as in the upper part of PK IV or PK VIII, the latter with a well expressed carbonate nodule Ck horizon, are probably formed synsedimentarily by simultaneous soil formation and loess accumulation. The lower Bt horizon of PK I is at least as strong developed as the Holocene soil because of considerable pedogenic clay formation (Fig. 3). Therefore this soil probably represents the *last interglacial*, equivalent to the 5e-stage in the δ^{18}O-record. TL-dating to check this is in progress. In all probability there was a *xeric* soil moisture regime in the interglacials during the Pleistocene. However almost all palaeosols were truncated and have lost their A or mollic A horizons, so we cannot decide whether they were Xerolls or Xeralfs – Phaeozems or Luvisols respectively. The only examples of good preservation are the upper parts of PK III and PK V (except the overlying CB horizon, see Fig. 4). Only these soils show an A horizon with a fine spongy fabric rich in aggregates and pores, which qualifies them as former mollic epipedons.

Conclusions and Discussion

As pointed out in section 1, the Brunhes/Matuyama boundary was found in many loess exposures in Central Asia between PK IX and PK X. In addition some exposures including Karamaydan and Chashmanigar (*e.g.* Dodonov 1991: Fig. 4; Lazarenko

1984: Fig. 13.1) show a striking angular unconformity below PK X mentioned above (section 2) which confirms the stratigraphic correlation of the palaeosols in different loess profiles. Assuming that the lower Bt horizon of PK I (Fig. 3) represents the last interglacial equivalent to stage 5e of the $\delta^{18}O$-record, the upper part of the loess-palaeosol-sequence in Karamaydan (PK I-X) can very well be compared with the $\delta^{18}O$-record shown in the adjusted SPECMAP curve for the *Brunhes Chron* (Imbrie *et al.* 1984; cf. Figs. 3,4). This agreement is even better than in the well-known Potou section near Luochuan in the Central Loess Plateau in China (Bronger & Heinkele 1989a: Fig. 2). Below stage 11 and even more below stage 15 the oscillations in the SPECMAP curve are less pronounced (Fig. 4) than above. Because of this Berger and Wefer (1992: 529ff) concluded that extremely cold periods and also really warm ("extreme") interglacials – like the Holocene – occured only in the "Milankovic" period (0-620 ka). However, several B or Bt horizons, mainly in PK V and PK VI, corresponding with stages 13 and stage 15, are strongly developed and thicker than the Holocene soil. The corresponding S5 pedocomplex at Luochuan and in many other exposures in the Central Loess Plateau in China - mainly its most strongly developed upper soil – is the most pronounced pedocomplex in the whole Potou section showing a much stronger pedogenesis than the younger Pleistocene soils S4-S1 and the Holocene soil in this area (Bronger & Heinkele 1989a). The stratigraphical equivalent "F6" in the profile of Stari Slankamen in the Carpathian Basin (Bronger 1976; Bronger & Heinkele 1989 b) also is much more strongly developed than the Holocene soil in this region, indicating a probably warmer and moister climate than today. These questions need further discussion. The conclusion of Dodonov (1991: 188) and Lazarenko (1984: 125, Fig. 13.1) that their PK V or their "soil complex 3" both equivalent to our PK IV, corresponds with the last interglacial according to TL–dates seems questionable.

The loess-palaeosol-sequence of the *Matuyama Chron* – at least its main parts – at Chashmanigar (Fig. 5) is much more detailed than even the equivalent sequence at Luochuan (Bronger & Heinkele 1989a) and probably also the Baoji loess section west of Xian / China (Ding *et al.* 1993). Unfortunately this section has only been described from a geologic-stratigraphical point of view. The central and lower part of the Chashmanigar section (Fig. 5) is, from our present knowledge, the most detailed loess-palaeosol-sequence in Central Asia. It provides more palaeoclimatic information regarding cold arid stages represented by loesses and warm humid stages (interglacials) represented by palaeosols than the loess profiles in China and even from the deep-sea cores known so far. The conclusion of Berger and Wefer (1992, 549) that the cycles of much less developed cold periods and interglacials especially in the Laplace period (between about 2 Ma and 1.2 Ma B.P.) cover only 41 ka – in comparison with a 100 ka cycle in the Milankovic period – is not in agreement with our results: most palaeosols in this period, separated by typical loesses (Fig. 5), not only the pedocomplexes XIX and XXV, are more strongly developed than the Holocene soil and represent longer interglacial periods.

Acknowledgements

The authors are very grateful to Prof. S. P. Lomov for demonstrating to us the Holocene soils near Karatau typical for this area and for stimulating discussions in summer 1991 and also the Tadjik Geological Survey for the help in getting the different parts of the Karamaydan section prepared in summer 1992. We thank the Deutsche Forschungsgemeinschaft for supporting field, micromorphological and laboratory work (Grant Br 303/26-1). Our sincere thanks go to J.A. Catt, Rothamsted Exp. Station, U.K., for improving our paper and an unknown reviewer for his valuable comments.

References

BERGER, W.H. & WEFER, G. (1992). Klimageschichte aus Tiefseesedimenten - Neues vom Ontong-Java-Plateau (Westpazifik). *Naturwissenschaften*, 79, 541-550.

BREWER, R. (1964). *Fabric and Mineral Analysis of Soils.* John Wiley & Sons, New York, 470pp.

BRONGER, A. (1966). Lösse, ihre Verbraunungszonen und fossilen Böden - ein Beitrag zur Gliederung des oberen Pleistozäns von Südbaden. *Schriften d. Geogr. Inst. d. Universität Kiel*, 25(2), 113pp.

BRONGER, A. (1969/70). Zur Mikromorphologie und zum Tonmineralbestand quartärer Lößböden in Südbaden. *Geoderma*, 3, 281-320.

BRONGER, A. (1974). Zur postpedogenen Veränderung bodenchemischer Kenndaten insbesondere von pedogenen Eisenoxiden in fossilen Lößböden. *In: Transactions of the 10th International Congress of Soil Science*, VI(II), Moscow, 429-441.

BRONGER, A. (1976). Zur quartären Klima- und Landschaftsgeschichte des Karpatenbeckens auf paläopedologischer und bodengeographischer Grundlage. *Kieler Geographische Schriften*, 45, 268pp.

BRONGER, A. (1991). Argillic horizons in modern loess soils in an ustic soil moisture regime? Comparative studies in forest-steppe and steppe areas from Eastern Europe and the USA. *Advances in Soil Science*, 15, 41-90.

BRONGER, A. (1994). Chernozems. Kastanozems. Phaeozems. *In:* ISSS - ISRIC - FAO. *World Reference Base for Soil Recources (Draft)*, pp. 90-101. Wageningen/Rome.

BRONGER, A. & HEINKELE, TH. (1989a). Micromorphology and genesis of paleosols in the Luochuan loess section, China: Pedostratigraphical and environmental implications. *Geoderma*, 45, 123-143.

BRONGER, A. & HEINKELE, TH. (1989b). Palaeosol sequences as witnesses of Pleistocene climatic history. *In:* Bronger, A. and Catt, J. (eds), *Palaeopedology - Nature and Application of Palaeosols*, Catena Supplement 16, 163-186.

BRONGER, A., PANT, R.K., SINGHVI, A.K., ERMLICH, J. & HEINKELE, T. (1986). Zur Löss-Boden-Chronostratigraphie und Quartären Klimageschichte des Kashmir Valley (Indien). *Göttinger Geographische Abhandlungen*, 81, 89-103.

BRONGER, A., PANT, R.K. & SINGHVI, A.K. (1987). Pleistocene Climatic Changes and Landscape Evolution in the Kashmir Basin, India: Palaeopedologic and Chronostratigraphic Studies. *Quaternary Research*, 27, 167-181.

BRONGER, A., BRUHN-LOBIN, N. & HEINKELE, TH. (1994).

Micromorphology of Palaeosols - genetic and palaeoenvironmental deductions. Case studies from Central China, South India, NW-Morocco and the Great Plains of the USA. *Developments in Soil Science*, 22,187-206.

BULLOCK, P. & THOMPSON, M. L. (1985). Micromorphology of Alfisols. *In:* Douglas, L. A. & Thompson, M. L. (eds), *Soil Micromorphology and Soil Classification*, pp. 17-47. SSSA Special Publication No. 15, Madison/Wisc.

BULLOCK, P., FEDOROFF, N., JONGERIUS, A., STOOPS, G., TURSINA, T. & BABEL, U. (1985). Handbook for Soil Thin Section Description, Waine Research Publications, Wolverhampton.

CATT, J.A. (1988). Soils of the Pliocene-Pleistocene: do they distinguish types of interglacial? *Transactions of the Royal Society of London*, 318, 539-557.

CATT, J. A. (1989). Relict Properties in Soils of the Central and North-West European Temperate Region. *In:* Bronger, A. and Catt, J.A. (eds), *Palaeopedology: Nature and Application of Palaeosols*, Catena Supplement 16, 41-58.

CATT, J.A. (ed.) (1990). Paleopedology Manual. *Quaternary International*, 6, 1-95 .

CATT, J.A. (1995). Soils in eolian sequences as evidence of Quaternary climatic change: problems and possible solutions. *Quaternary Proceedings*, (this volume).

COURTY, M.A, DHIR, R. P. & RACHAVAN, H. (1987). Microfabrics of calcium carbonate accumulations in arid soils of Western India (Rajastan). *In:* Fedoroff, N. *et al.* (eds), *Soil Micromorphology*, pp. 227-234. AFES, Plaisir (France).

DING, Z., RUTTER, N. & LIU, T.S. (1993). Pedostratigraphy of Chinese Loess Deposits and Climatic Cycles in the past 2.5 Myr. *Catena*, 20, 73-92.

DODONOV, A.E. (1979). Stratigraphy of the Upper Pliocene-Quaternary Deposits of Tajikistan (Soviet Cental Asia). *Acta Geologica Academiae Scientiarum Hungaricae*, 22(1-4), 63-73.

DODONOV, A.E. (1984). Stratigraphy and Correlation of Upper Pliocene-Quaternary Deposits of Central Asia. *In:* Pecsi M. (ed.). *Lithology and Stratigraphy of Loess and Paleosols*, pp. 201-211 . Budapest.

DODONOV, A.E. (1991). Loess of Central Asia. *GeoJournal*, 24(2), 185-194.

DODONOV, A.E. & LOMOV, S.P. (1985). Stratigraphy and Pedogenesis of Loess Formation of Southern Tajikistan. *In: Current Trends in Geology Vol. VI* (Climate and Geology of Kashmir and Central Asia), 223-225.

DODONOV, A.E., MELAMED, Y.R. & NIKIFOROVA, V.V. (eds) (1977). International Symposium on the Neogen-Quaternary Boundary, Moscow, IGCP 41, *Excursion Guidebook*.

DODONOV, A.E., MAVLYANOV, G.A. &TETYUKHIN, G.F. (1982). *Guidebook for Excursions A-II & C - Uzbek SSR,*

Tajik SSR, INQUA XI Congress, Moscow.

DODONOV, A.E. & PENKOV, A.V. (1977). Some data on the stratigraphy of the watershed loesses in Tajik depression. *Bulletin of the Commission on the Quaternary Research*, 47, 67-76.

FAO - UNESCO (1974). Soil Map of the World, Vol. I (Legend). Paris.

FÖRSTER, TH. & HELLER, F. (1994). Palaeomagnetism of loess deposits from the Tajik depressiom (Central Asia). *Earth and Planetary Science Letters* (in print).

GUO, Z., FEDOROFF, N. & AN, Z. (1991). Genetic Types of the Holocene Soil and the Pleistocene Palaeosols in the Xifeng Loess Section in Central China. *In:* Liu, T.S. (ed.), *Loess, Environment and Global Change*, Beijing, Science Press, 93-111.

IMBRIE, J., HAYS, J.D., MARTINSON, D.G., McINTYRE, A., MIX, A.C., MORLEY J.J., PISAS, N.G., PRELL, W.L. & SHACKLETON, N.J. (1984). The orbital theory of Pleistocene Climate: support from a revised chronology of the marine $\delta^{18}O$ record. *In:* Berger, A.L., Imbrie, J., Hays, J., Kukla, G. and Saltzman, B. (eds), *Milankovitch and Climate*, Part 1, 269-305.

ISSS - ISRIC - FAO (1994). *World Reference Base for Soil Resources (Draft)*. Spaargaren, O. (Ed.) Wageningen/Rome. 161pp.

KUBIENA, W.L. (1959). Prinzipien und Methodik paläopedologischer Forschung im Dienste der Stratigraphie. *Zeitschrift deutsch. Geol. Ges.*, 111, 643-652.

KUBIENA, W.L. (1964). Zur Mikromorphologie und Mikromorphogenese der Lößböden Neuseelands. *In:* Jongerius, A. (ed.), *Soil Micromorphology*, pp. 219-235. Amsterdam.

LAZARENKO, A.A. (1977). Loess cover of the Tajik Depression (Stratigraphy, Lithology, Problems of Genesis). *In: Abstracts of the IGCP International Symposium on the Neogene-Quaternary Boundary*, pp. 35-37. Moscow.

LAZARENKO, A.A. (1984). The Loess of Central Asia. *In:* Velichko (ed.). *Late Quaternary Environments of the Soviet Union*, pp. 125-131. University of Minnesota Press, Minneapolis.

LAZARENKO, A.A., BOLIKHOVSKAYA, N.S. & SEMENOV, V.V. (1981). An attempt at a detailed stratigraphic subdivision of the loess association of the Tashkent region. *Internat. Geology Rev.*, 23(11), 1335-1346.

LAZARENKO, A.A., SEMENOV, V.V. & PENKOV, A.V. (1991). Excurses of geomagnetic field during the Brunhes Epoch in loess formation of Central Asia and their stratigraphical significance. *In: Abstracts INQUA XIII. Inter. Congr. Beijing*, 187.

LOMOV, S.P. (1985). Soils of the major typical landscapes of Hissar nature economic field. Tadjik. Academy of Science, Dushanbe (Donish Publishers).

LOMOV, S.P. & RANOV, V.A. (1985). The Peculiarities of the Pleistocene Palaeosol Formations and Distribution of Embedded Palaeolithic Tools. *In: Current Trends in Geology, Vol. VI* (Climate and Geology of Kashmir and Central Asia), 227-240.

LOMOV, S.P. & SOSNOVSKAYA, V.P. (1977). Geographical and Genetic Characteristics of the Brown Mountain-Forest Soils of Tadzhikistan. *Soviet Soil Science*, 9(4), 1-13.

McKEAGUE, J. A. (1983). Clay Skins and Argillic Horizons. *In:* Bullock, P. and Murphy, C. P. (eds), *Soil Micromorphology*, Vol.II Soil Genesis, pp. 367- 387. AB Academic Publishers, Berkhamsted.

MÜLLER, M.J. (1980). Handbuch ausgewählter Klimastationen der Erde. Trier: Forschungsstelle Bodenerosion der Universität Trier.

PENKOV, A.V. & GAMOV, L.N. (1977). Paleomagnetic Datums in Pliocene-Quaternary Strata of Southern Tajikistan. *In: Abstracts of the IGCP International Symposium on the Neogene-Quaternary Boundary*, pp. 46-47. Moscow.

PENKOV, A.V. & GAMOV, L.M. (1980). Palaeomagnetic datums in the Pliocene to Quaternary Strata of Southern Tajikistan. *In: The Neogene-Quaternary Boundary*, (IGCP Projekt No. 41), pp. 189-194. Nauka, Moscow.

SCHROEDER, D. (1984). Soils - Facts and Concepts. Bern, International Potash Institute, 140pp.

SMOLIKOVA, L. (1967). Polygenese der fossilen Lößböden der Tschechoslowakei im Lichte mikromorphologischer Untersuchungen. *Geoderma*, 1, 315-324.

SMOLIKOVA, L. (1971). Gesetzmäßigkeiten der Bodenentwicklung im Quartär. *Eiszeitalter u. Gegenwart*, 22, 156-177.

SMOLIKOVA, L. (1990). Problematika paleopedologie. Regionalni paleopedologie. *In:* Nemecek, J., Smolikova, L. and Kutilek, M. (eds): *Pedologie a paleopedologie*, pp. 381-479. Praha.

SOIL SURVEY STAFF (1975). Soil Taxonomy. A Basic System of Soil Classification for Making and Interpreting Soil Surveys. Agriculture Handbook No. 436, Washington D.C., U.S. Govt. Printing Office, U.S.D.A. Soil Conservation Service.

STANINKOVITCH, K. (1968). Geobotanic zonation of Tadjikistan. *In: Atlas of Tajik SSR*. Academy of Sciences of Tajik SSR, Moscow, Leningrad, (russ.).

Van WAMBEKE, A., HASTINGS, P. & TOLOMEO, M. (1986). *Newhall Simulation Model*: A Basic Programm for the IBM PC (incl. manual, 37 pp.). Department of Agronomy, Cornell University, Ithaca, New York.

WIEDER, M. & YAALON, D.H. (1982). Micromorphological Fabrics and Developmental Stages of Carbonate Nodular Forms related to Soil Characteristics. *Geoderma*, 28, 203-220.

WILDING, L.P., SMECK, N.E. & HALL, G.F. (Eds.) (1983). Pedogenesis and Soil Taxonomy - II. Soil Orders, Amsterdam: Elsevier.

ZACHARIAE, G. (1964). Welche Bedeutung haben Enchytraeen im Waldboden? *In:* Jongerius, A. (ed.), *Soil Micromorphology*, pp. 57-68, Elsevier, Amsterdam.

ZAPRIAGAEVA, V.I. (1964). *Wildgrowing Fruits of Tajikistan*. Nauka Publ. House, Moscow, Leningrad, 695 pp, (russ.).

ZAPRIAGAEVA, V.I. (1968). *Map of wildgrowing fruits*, 1:1,5 Mill. *In: Atlas of Tajik SSR*, Academy of Sciences of Tajik SSR, Moscow, L,eningrad, (russ.).

ZAPRIAGAEVA, V.I. & IKONIKOV, V. (1968). *Map of forests and distribution of rare trees*, 1:1,5 Mill. *In: Atlas of Tajik SSR*, Academy of Sciences of Tajik SSR, Moscow, Leningrad, (russ.).

Quaternary Proceedings No. 4, 1995 83-93

Weathering and Pedogenesis of Wind-Blown Sediments in the Mount Carmel Caves, Israel.

A. Tsatskin, M. Weinstein-Evron and A. Ronen.

Tsatskin, A., Weinstein-Evron, M. and Ronen, A., 1995 Weathering and pedogenesis of wind-blown sediments in the Mount Carmel caves, Israel, In *Wind Blown Sediments in the Quaternary Record* (Edward Derbyshire). Quaternary Proceedings No. 4, John Wiley & Sons Ltd., Chichester, pp. 83-93.

Abstract

Current sedimentological studies of the Tabun and Jamal caves, Mount Carmel, Israel, are focused on the portion of their sedimentary fill containing Lower Palaeolithic lithic assemblages. These complex sedimentary sequences show signs of erosion and karstic collapse and a diversity of pedo-geochemical transformations. Wind-blown sands and silts, better preserved in Tabun, were probably intermixed with other materials and largely reworked by various post-depositional processes. A range of processes is outlined: low-energy run-off and colluviation, chemical weathering of both mineral mass and bones, movement and redeposition of soluble products, anthropogenic and biological activity. Studies of petrographic thin sections suggest that disrupted charcoal fragments, comminuted organic materials and chips of bones with signs of thermal processing are still preserved in the sediments altered through diagenesis. SEM/EXDRA studies confirm strong weathering of the mineral mass, and show that various alterations of bone hydroxyapatite resulted in the formation of transluscent coatings, pseudomorphs and nodules. It is still controversial whether transformation of bones occurred penecontemporaneously or in later diagenesis, during inundation and ground water impact in the caves. Phosphatic pseudomorphs are related sometimes to biological activity features; they show a complex polyphase internal fabric, and nonuniform distribution of clay, phosphates, iron and other elements. It is tentatively suggested that Fe and Ti were released from silicates and oxides by acid (guano-derived ?) weathering, and then attached to phosphatic pseudomorphs and nodules. Breakdown of bones was undoubtedly enhanced during ground water rise, which is testified by the more advanced mineralogical and geochemical transformations in specific geomorphic situations.

KEYWORDS: wind-blown sediments, prehistoric caves, micromorphology, post-depositional changes, phosphatization.

Tsatskin, A., Weinstein-Evron, M. and Ronen, A., Zinman Institute of Archaeology, University of Haifa, Haifa 31905, Israel.

Introduction

Studies of sediments from prehistoric cave sites are a significant part of Quaternary research. Wind-blown sediments in caves are usually intermixed or interbedded with colluvial, spelaean and anthropogenic materials, which all underwent intense weathering (Laville *et al.* 1980; de Lumley *et al.* 1981; Weiner & Bar-Yosef 1990). Courty *et al.* (1989) emphasized the important role of soil micromorphology techniques for identification of depositional and post-depositional features in cave sequences. Application of these techniques together with geochemical and mineralogical studies in Middle Palaeolithic caves in the Southern Levant has furnished many important details about the Late Pleistocene history of the area (*e.g.* Bar-Yosef *et al.* 1992; Weiner *et al.* 1993). Much less is known about processes of sedimentation and weathering in earlier, Lower Palaeolithic cave sites.

Such early sediments are found in Tabun and Jamal caves, in Nahal Me'arot (Wadi Mughara), Mount Carmel (Fig. 1). The caves are situated on the southern bank of the wadi, on the western cliff of Mount Carmel, 20 km south of Haifa, facing the Mediterranean Coastal Plain at 45-60 m amsl. Garrod was the first to describe the long and rich archaeological sequence in Tabun (Garrod & Bate 1937), extending from the Lower

Palaeolithic (layers G to E) to the Middle Palaeolithic (layers D to B) (Fig. 2). In the renewed excavations of the Lower and Middle Palaeolithic in the late 1960's and early 1970's (Jelinek *et al.* 1973), researchers placed greater emphasis on palaeoenvironmental and sedimentological aspects. Since the 1980's Tabun excavations have been primarily in the Lower Palaeolithic parts (Ronen 1991, 1993).

Goldberg (1973) and Farrand (1979) distinguished three depositional units in the cave and related them to the previous subdivisions by Garrod (Fig. 2). Goldberg (1973), who examined mainly granulometric, chemical, and mineralogical properties of the various layers in Tabun sequence, has stated that the Lower Palaeolithic sediments of Unit III are essentially wind blown, ranging from silty sands at the bottom to sandy silts at the top. Analyses by scanning electron microscope (SEM) showed rounded, smoothed and pitted surfaces of the fine sand grains, characterisitic of dune sands. He further argued that the amount of unstable heavy minerals, such as hornblende and epidote, is significantly less in the lowermost unit III than in the overlying units, whereas stable minerals (zircon, tourmaline and rutile) are equally preserved. From this it was inferred that greater weathering took place in Unit III of Tabun Cave. Unit III has been severely disturbed by karstic collapse, which caused slumping of the lowermost layers G and F into a

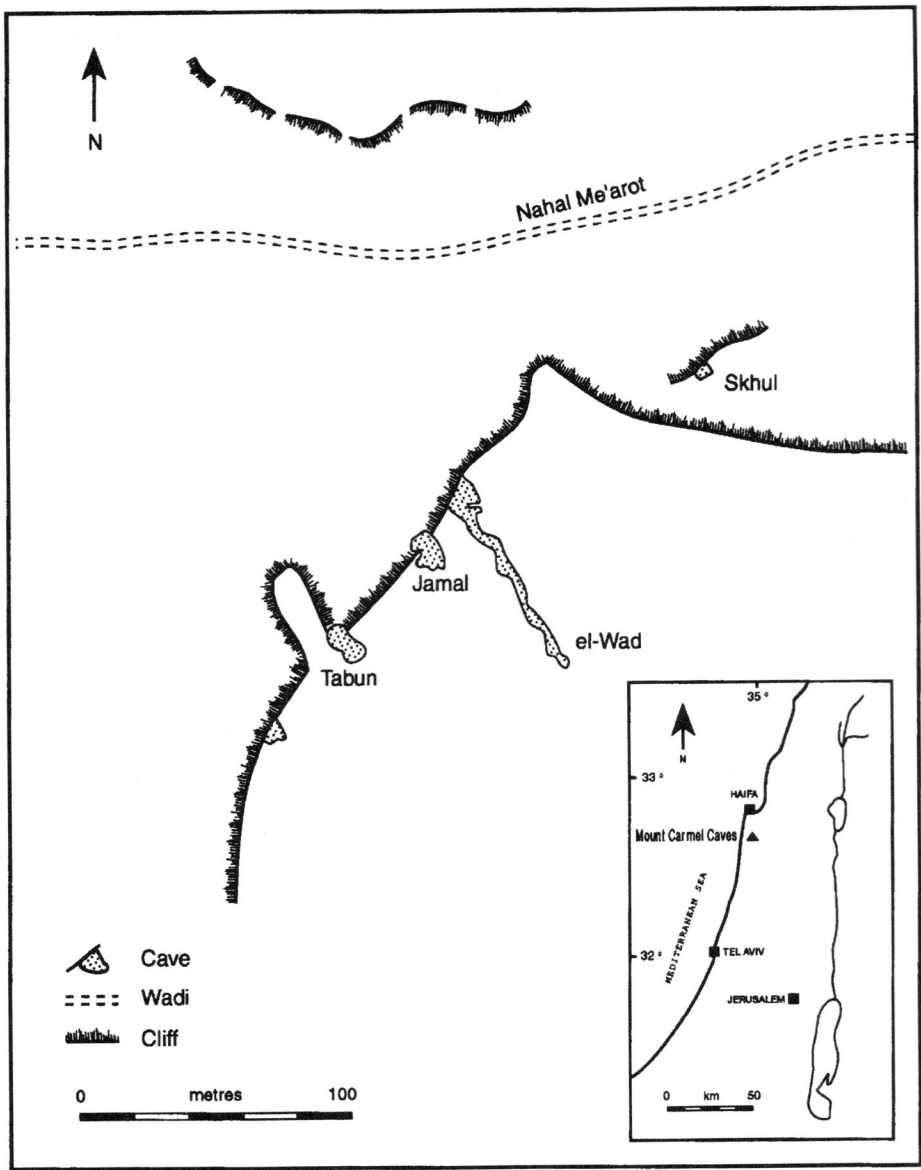

Figure 1.   Map of Mount Carmel caves.

sinkhole in the centre of the outer chamber. Lateral disconformities could have resulted also from the action of an ancient spring (Fig. 2), the sediments of which were recorded in previous excavations (Garrod & Bate 1937) but are no longer preserved. In the course of ground water rise the wind-transported sediments underwent strong weathering and phosphatisation (Goldberg & Nathan 1975).

A geoarchaeological study in the neighbouring, newly excavated Jamal Cave (Fig. 1) is still in progress (Weinstein-Evron & Tsatskin, nd1,2 in press). Though reduced in thickness, strongly tilted and truncated by erosion, the sequence can be divided into three lithological units (Fig.3). Units 2 and 3 contain Lower Palaeolithic (Acheuleo-Yabrudian ?) implements. Unit 1 is of Middle Palaeolithic affiliation (Weinstein-Evron & Tsatskin, Nd1 in press). The Jamal sequence can thus be roughly correlated with that of Tabun. The sediments of Jamal Cave are hard, strongly brecciated and contain large amounts of clay. These sediments possibly accumulated from different sources, both aeolian and colluvial, indicating complex sedimentation processes and post depositional changes.

The chronological framework of the Lower Palaeolithic in Mount Carmel caves has been substantially revised recently.

Jelinek et al. (1973) related the entire sequence of Tabun cave to the Upper Pleistocene based on the sedimentological data and geomorphological correlations with marine terraces in the Eastern Mediterranean coast. The Lower Palaeolithic was dated in these ways to the Riss-Würm Interglacial or to oceanic oxygen isotope stage 5 (Jelinek 1982). The overlying Middle Palaeolithic was put primarily within isotope stage 4, as suggested by the then available C14 dates (Table 1). But a greater age was suggested from biostratigraphical and geological considerations (Tchernov 1981; Bar-Yosef & Vandermeersch 1981). ESR and TL dates for sites in the Levant, including Skhul (Stringer et al. 1989), Kebara (Valladas et al. 1987) and Qafzeh (Schwarcz et al. 1988) suggest that the Middle Palaeolithic dates from isotope stage 5. Recent TL and ESR dates for Tabun (Table 1) imply an even older age for both the Middle and the Lower Palaeolithic layers in the cave (Grun et al. 1991; Mercier & Valladas 1994; see also McDermott et al. 1993 and Porat et al., nd, in press). According to these new dates, much of the Tabun sequence should be placed within the Middle Pleistocene.

This paper describes the accumulation, redistribution, erosion, and secondary geochemical transformation of Middle

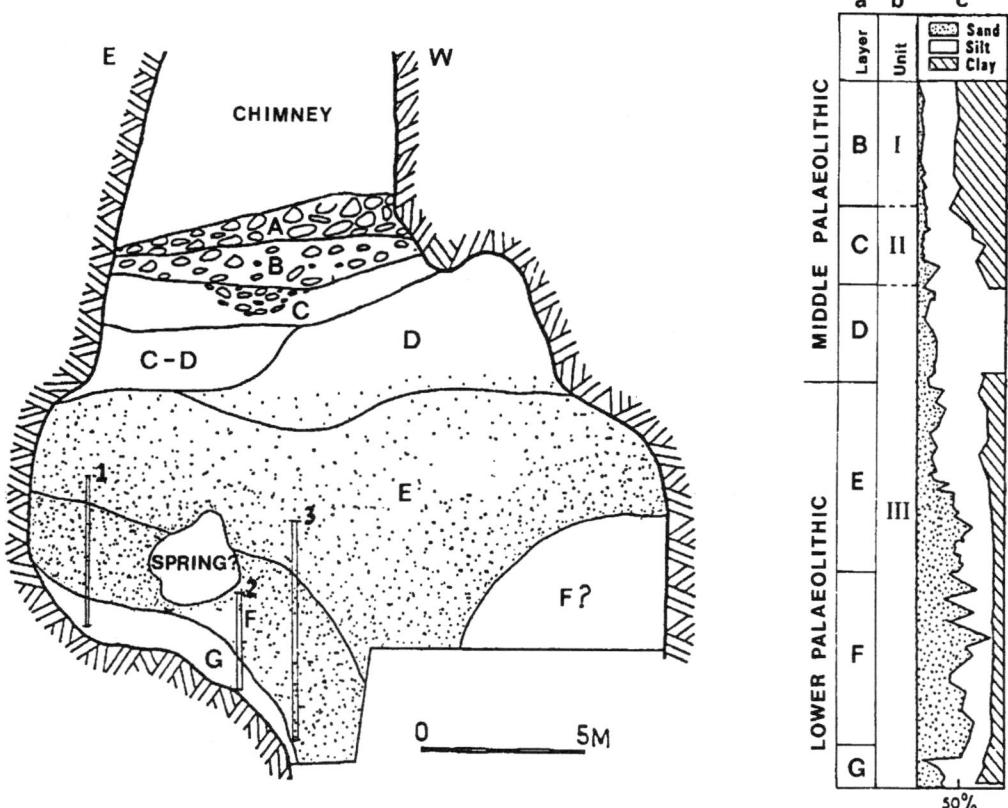

Figure 2.　Tabun Cave; left – Schematic representation of archaeological layers (after Garrod & Bate 1937). 1,2,3 refer to the sections which were analyzed for the current work; right – Sedimentology (after Goldberg 1973): a – Garrod's layers; b – the main sedimentological units; c – changes in granulometry.

Table 1: C14, ESR and TL dates from sites mentioned in the text.

		DATING METHOD			
SITE	**C14[1]**	**ESR[2] (in ka)**		**TL[3] (in ka)**	
		EU	**LU**		
Tabun					
B	39,700±800 (GrN-2534)	86±11	103±16		
C	40,900±1000 (GrN-2729)	102±17	119±11	134/184±29 195/263±25	
C-D				226/307±30	
D		122±20	166±20	249/297±57	
Ea		154±34	188±31	202/270±23	
Eb		151±21	168±15	229/296±18	
Ec		176±10	199±7	273/350±37	
Ed		182±61	213±46	331±30	
Qafzeh	96±13	115±15	92±5		
Skhul		81±15	101±12	113±18	
Kebara					
X		60±5.9	64±5.5	61±3.6	
XII				60±3.5	

EU = Early Uptake;　　　　LU = Linear Uptake

1.　Vogel and Waterbolk, 1963.　　　2.　Grün *et al.*, 1991.　　　3.　Mercier and Valladas, 1994.

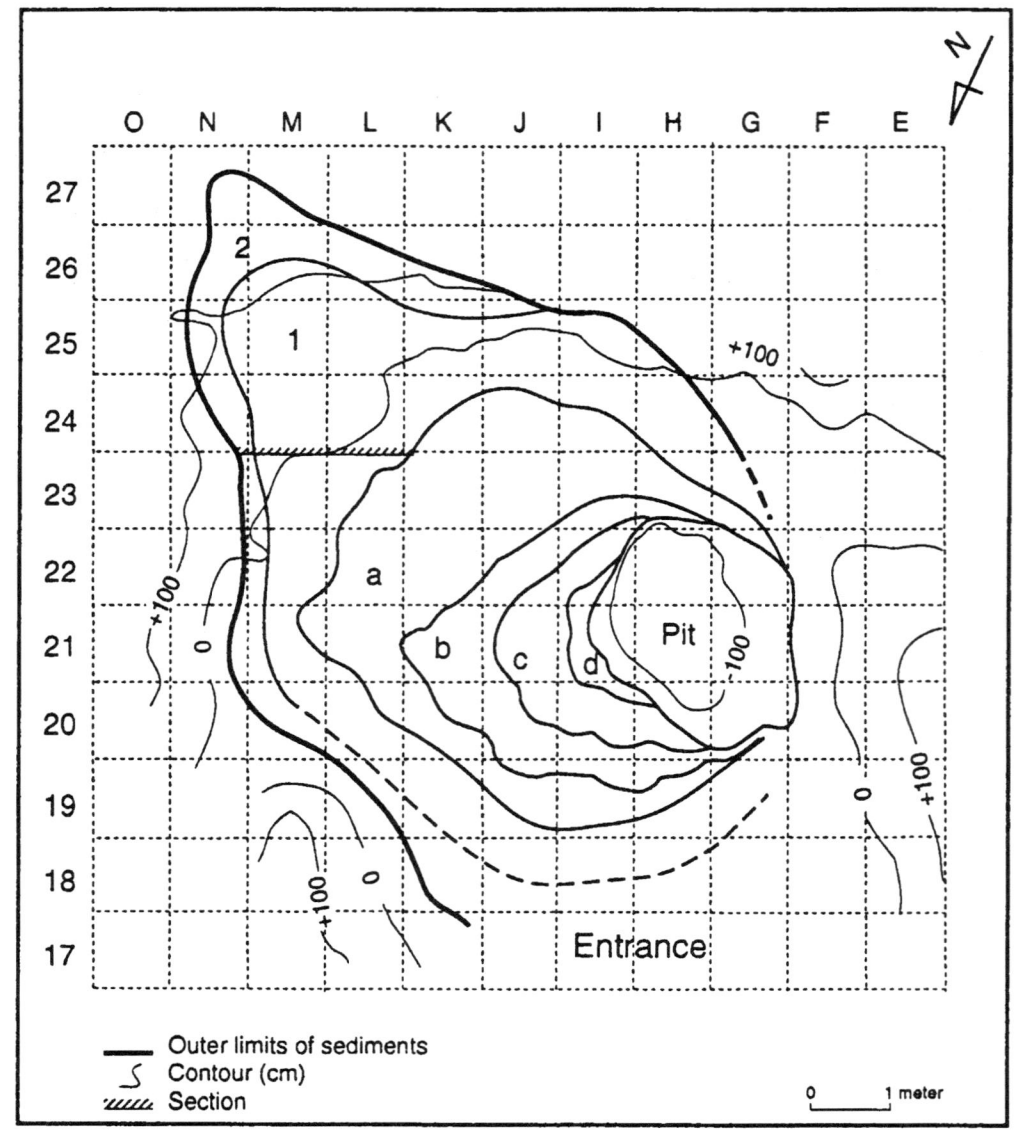

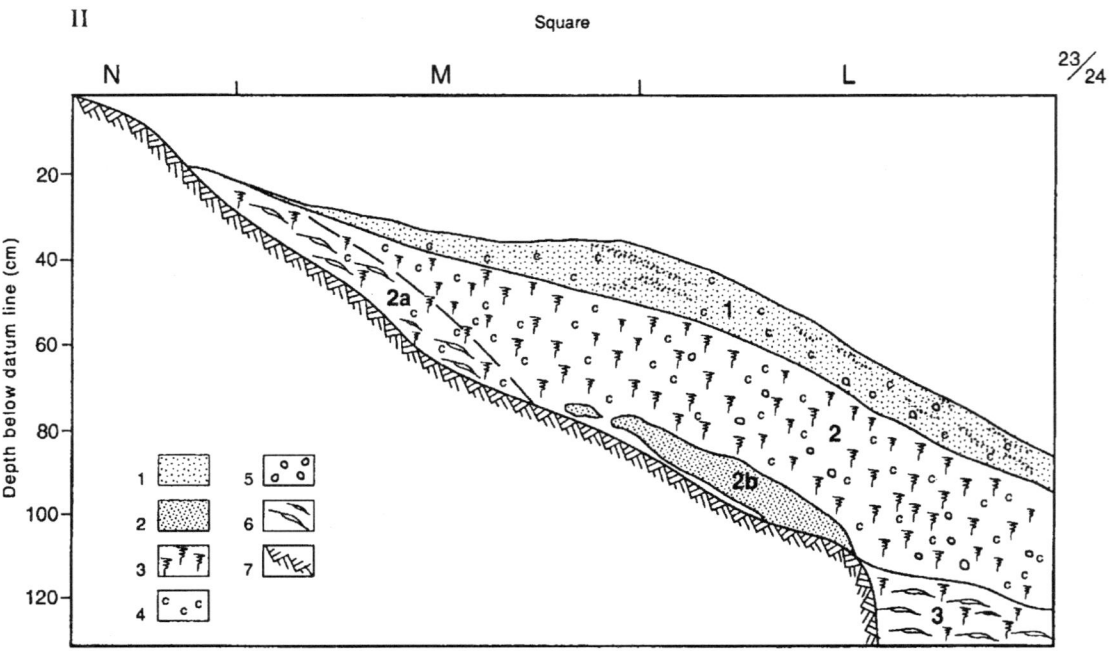

Figure 3.   Jamal Cave. I – Ground plan of the cave showing the concentric arrangement of the main, tilting geological units; II – The southern wall of the east-west section along line 23. For location of the section see Fig. 3, I. 1 – silty clay loam; 2 – phosphatised clay loam; 3 – reddish clay loam; 4 – CaCO$_3$ concretions; 5 – phosphatic mottles; 6 – gleying features; 7 – bedrock.

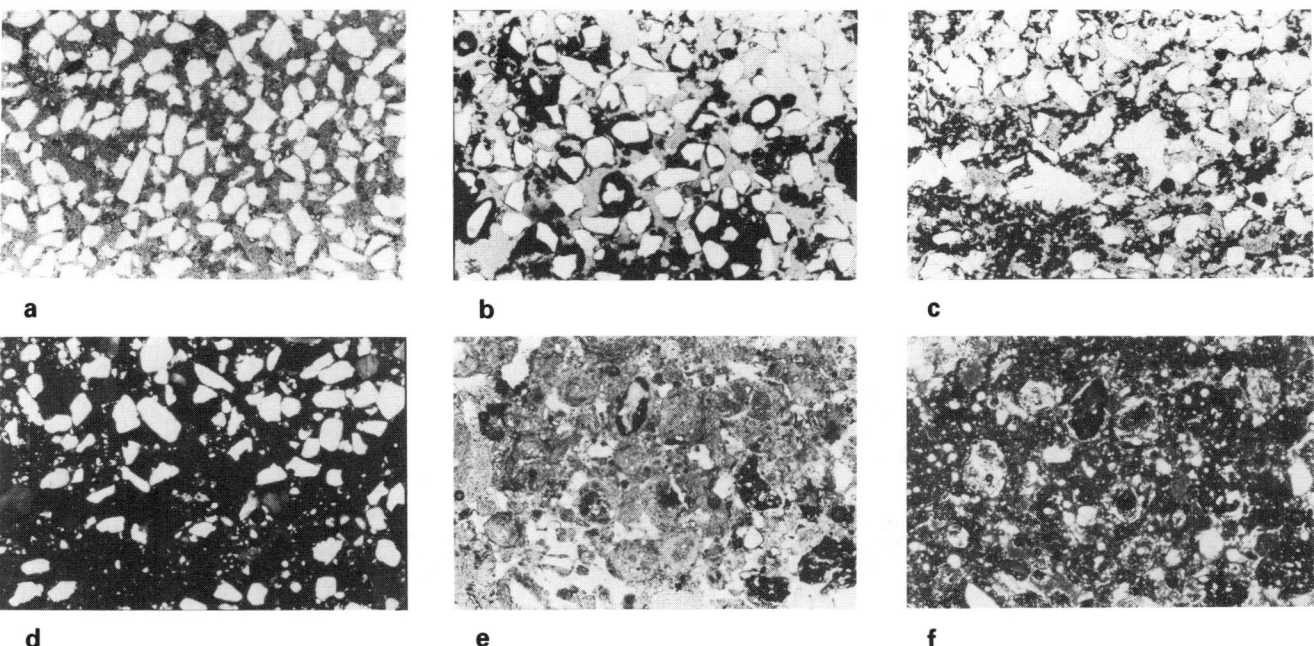

Figure 4. Microfabric of the aeolian and colluvially reworked sediments in Tabun and Jamal caves (the frame length is 2,8 mm). (a) The best preserved wind-blown sediments at the bottom of layer E in Tabun (around the sinkhole). Moderately well-sorted poorly cemented subangular fine sands; plane polarized light (PPL). (b) Wind-blown sands covered by clay coatings with signs of secondary biogenic and mechanical reworking, layer F, Tabun, PPL. (c) Signs of low-energy run-off in layer E, Tabun: Alternation of sandy laminae (upper part of the photo) with silty-sand laminae, enriched with isotrophic phosphatic cement (PPL). (d) The same view taken in crossed polarised light (XPL). (e) Colluvially reworked phosphatised silty clay sediments of Jamal, Unit 1: Heterogeneous ooid-like microfabric with 0.25 mm ooids (PPL). (f) The same view in XPL; note b-fabric (high birefringence) of particular ooids.

Pleistocene sediments containing Lower Palaeolithic assemblages in Tabun and Jamal caves, using various microscopic techniques.

Methods

This study is based on field observations of Tabun and Jamal, and systematic analyses using a polarising optical microscope. Samples for micromorphological studies in Tabun were collected from three areas in the cave (Fig. 2): near the NE wall, the 'spring' area and the sinkhole (1,2,3 respectively). The description of petrographic thin sections follows primarily the terminology of Bullock *et al.* (1985). A Jeol 840 scanning electron microscope (SEM) connected with a microprobe analyzer (EXDRA) was used to characterize the internal fabric and chemical composition of particular features. All SEM photographs present back-scattered electron images.

Results

The following sections contain an overall discussion of processes, rather than a description of individual samples, which is beyond the scope of this paper.

1. Primary Materials and Signs of their Mechanical Redistribution

In Tabun, the aeolian sediments are best preserved in the middle part of Unit III, Garrod's layer E (Fig. 2). They appear as pale yellow, bedded, silty sands, sporadically compacted by phosphates. In the outer chamber of the cave, loess-like sediments of seemingly similar age are apparently cross-bedded, which is characteristic of aeolian dunes on the adjacent Coastal Plain. In thin sections the wind-blown sediments in Tabun are represented by patches of fine sands composed of quartz, feldspar and small amount of heavy minerals. Well to moderately sorted subangular fine sand grains are embedded in the pale-green isotropic, phosphate rich groundmass (Fig. 4a). The presence of coated sand grains (Fig. 4b) is tentatively interpreted as an indication of saltation from the nearby exposures. A common feature in Nahal Me'arot caves is the evidence for intermittent redeposition of allochtonous material throughout the geological history of the sites. In Tabun Cave, the intensity of colluvial reworking varies within different depositional facies: the least reworked by slope movement are the sediments adhering to the walls of the cave, and the most reworked are those close to the sinkhole (Fig. 2). Here, in the central part of the cave, bedding patterns are most clearly expressed. Thin sections also show the alternation of microlaminae of coarse and fine-grained materials (Fig. 4c,d), indicating processes of low-energy run-off in the course of deposition.

In Jamal cave micromorphological features of secondary redeposition are expressed as partially welded ooid microaggregates of heterogeneous units (Fig. 4e,f). Ooids probably originated through the transportation and agitation of aggregates under conditions of repeated cycles of water saturation and desiccation (Mücher & Morozova 1983).

2. Post-Depositional Alterations of Cave Sediments

The Lower Palaeolithic sediments of Tabun and Jamal caves exhibit numerous post-depositional, biogeochemical changes, which occurred over a long time span. Some of the transformations can be related to anthropogenic activity, but others could have occurred later, following ground water intrusion into the cave system.

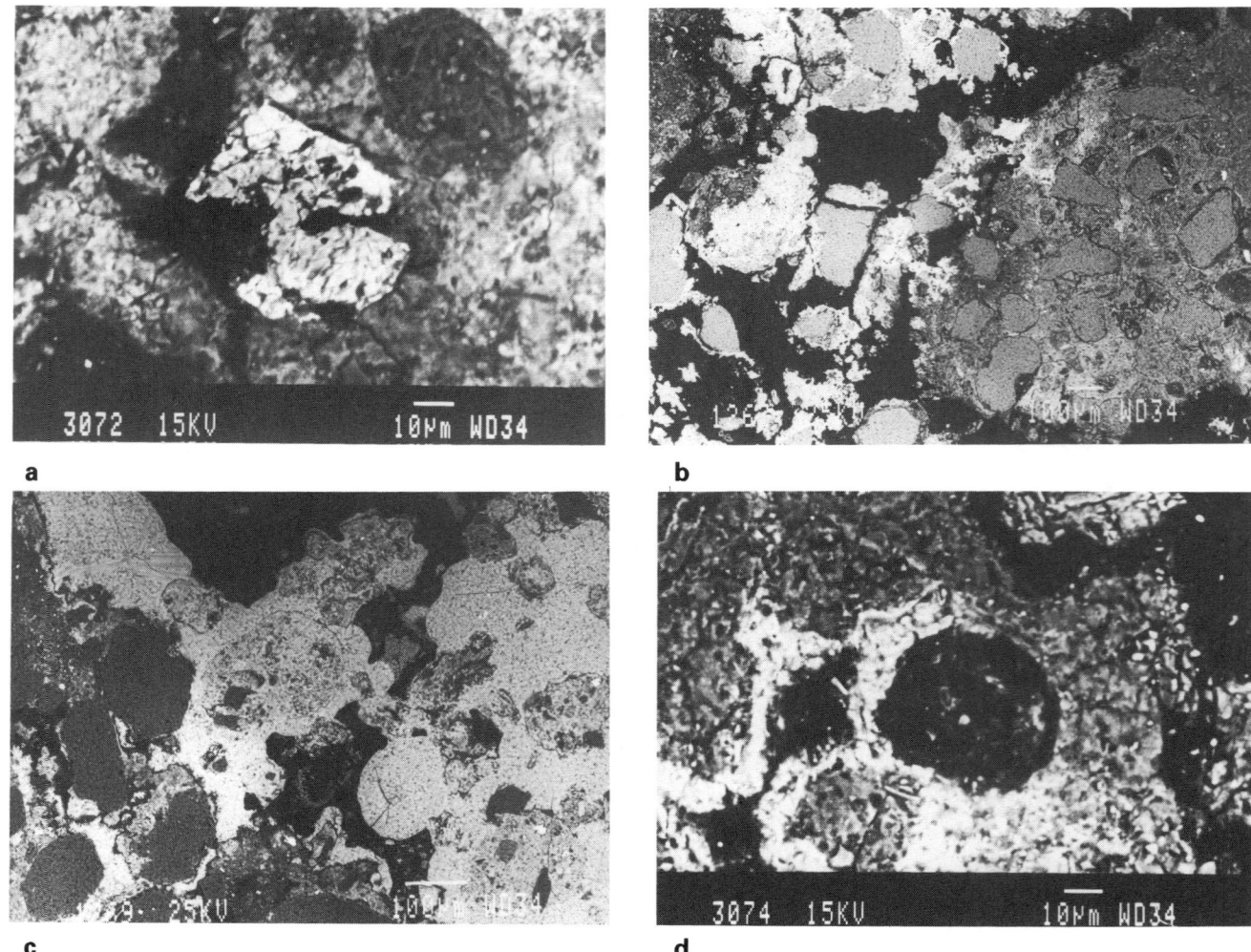

Figure 5.   SEM photographs, back-scattered electron image (BSE-image) of Tabun and Jamal sediments. (a) Silt-size ore-mineral grain, probably decayed ilmenite (bright) embedded in a dense clayey matrix; Unit a, Jamal Cave. (b) Rounded large bioaggregate composed of subangular and subrounded quartz grains embedded in clayey matrix (intermediate grey); note thin films surrounding individual grains, thicker in embayments (dark); a large void is partially filled with micritic calcite (light); transition between layers F and E, Tabun, near the wall. (c) Phosphatic coating (light) on the walls of a large void, bottom of layer E, Tabun, near the wall; note radial internal fabric in certain points. (d) Manganese coatings (light) surround OM-rich (?) clay clumps (black) against the clayey groundmass of Unit a, Jamal Cave.

a. Processes of weathering of mineral mass

Our SEM-EXDRA analyses show that silt grains of ore minerals (Fig. 5a), probably ilmenite, in both Tabun and Jamal, are strongly corroded and weathered. Ore minerals were also observed in thin sections. Weathering of ilmenite and other Fe- and Ti-bearing minerals (*e.g.* hornblende) mobilised iron and titanium, probably in acid (guano-derived?) solutions. Microprobe analysis (Fig. 6a) demonstrates a complex mixture of Ti, Fe, P, Ca, Si and Al, suggesting that Ti,Fe- phase was eventually redeposited, together with clay and phosphate minerals, on the surface of primary mineral grains. Sometimes the presence of ilmenite is identified by SEM-EXDRA analyses within particular microaggregates (Fig. 5b, 6b). These data suggest that the iron mineralisation in the caves results from *in situ* weathering and are consistent with previous observations by Goldberg (1973). Other sources of iron could have been either inwashing of terra rossa soil material or transportation in ground water from volcanic sources, as reported in the vicinity of Nahal Me'arot caves (Ilani 1989).

b. Phosphatisation.

In the Lower Palaeolithic sediments of Nahal Me'arot caves,

bones are usually not preserved. The lowermost sediments of Tabun contain many orange-yellow to dark green, 0.7 to 3-6 cm across, concretions. Unpermeable, indurated crusts are also observed, in particular in the central part of the cave, around the swallow hole. In Jamal, phosphates form compacted patches interspersed with calcite and manganese materials. Thin sections analyses show abundant small bone fragments in various state of preservation: from chips with their original structure (Fig. 7a,b,c) to bone pseudomorphs which take the form of nodules with complex internal fabric (Fig. 7d). Assimilation of phosphorus (from bones or faeces) into the clayey mass gives isotropic phosphate-impregnated matrix (Goldberg & Macphail 1990). This is the case in Tabun. In Jamal the impregnation increased the birefringence of the groundmass, apparently in the course of intense erosion of fine-grained materials.

Another common phosphatic feature observed in thin sections is the translucent coating which surrounds aggregates or fills pores. At low magnifications these usually take the form of smooth, laminated coatings (Fig. 7e,f), probably precipitated from solutions migrating through the pores. By SEM analysis, the coatings appear as coalescing spherical granules with filament-like radial internal fabric (Fig. 5c). EXDRA confirms that they are composed of pure hydroapatite (Fig. 6c). This

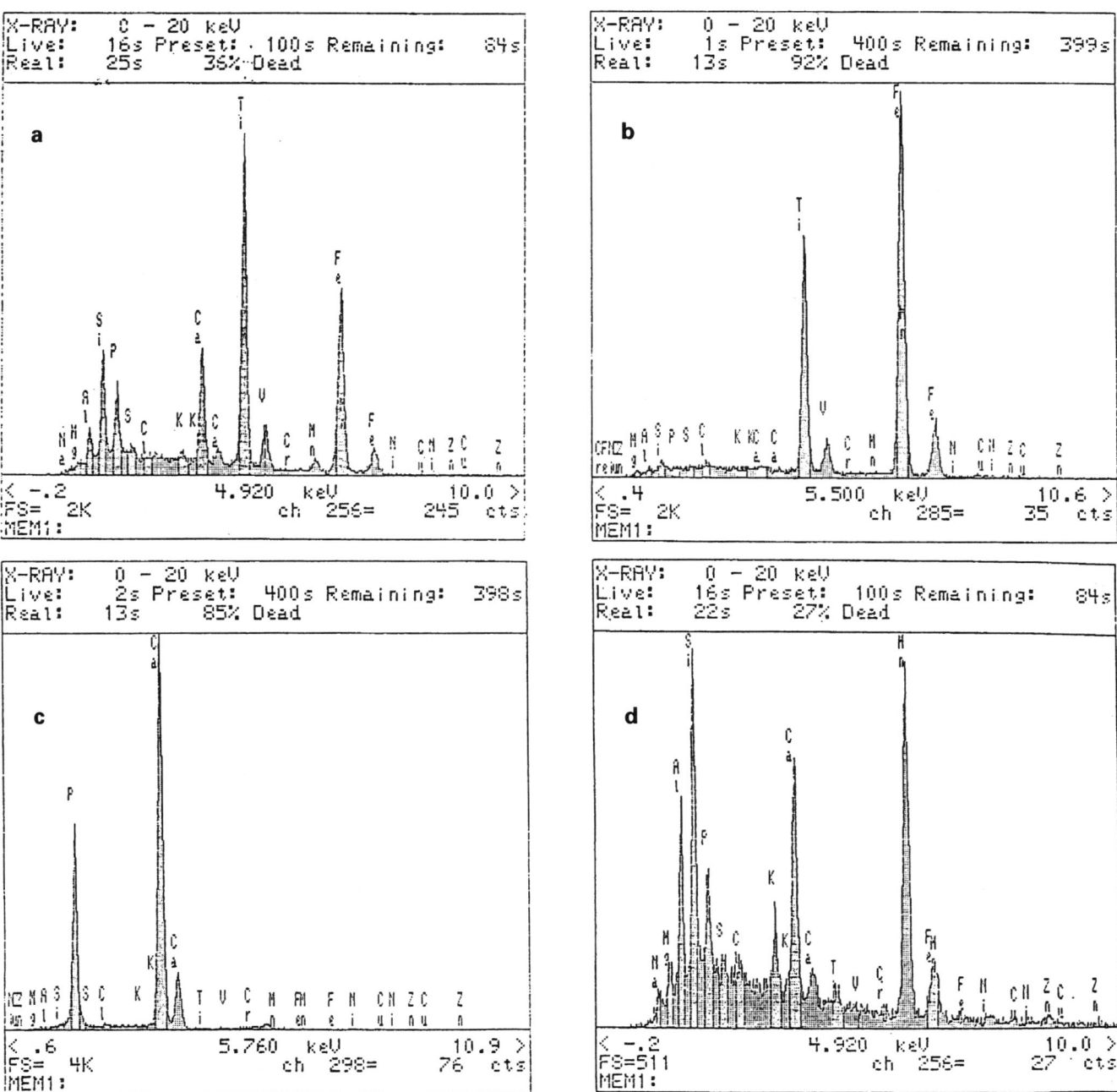

Figure 6. Chemical composition (EXDRA) of particular features demonstrated on Fig. 5. (a) Decayed ilmenite, Jamal Cave (Fig. 5, a); note the approximately equal amounts of Fe, Si, Ca, P against the Ti peak. (b) Bioaggregate from F to E transition, Tabun Cave (Fig. 5, b); the light spots in the clay matrix are composed of Fe and Ti (silt-size ilmenite). (c) Hydroapatite coating composed of P and Ca (Fig, 5, c). (d) Manganese coatings in Unit 1, Jamal Cave (Fig. 5, d); note a variety of elements with peaks in Mn and Si.

particular morphology of the hydroapatite coatings suggests that microbiological activity could have also been involved in their formation.

SEM/EXDRA studies show that nodules and pseudomorphs differ by their chemical composition from hydroapatite coatings in that phosphates within them are usually associated with ferric oxides and/or clay minerals. For example, the groundmass within a large bioaggregate at the bottom of layer E in Tabun (Fig. 5b) consists of an intimate mixture of quartz clay, phosphate and iron oxides; silt size titanium bearing minerals are also present (Fig. 6b). SEM/EXDRA analyses suggest that the nodules could have originated from replacement of apatite in the sediment by Al-Fe phosphates, after complete removal of calcite by groundwater (Peneaud 1978,1981; Flicoteaux & Lucas 1984).

c. Calcification

The Lower Palaeolithic sediments in Tabun and Jamal are completely leached of detrital calcite. Secondary calcite is absent from the central part of Tabun, around the ancient sinkhole, but quite common in other depositional facies of the cave. The accumulation of carbonates reaches its peak in Jamal (Fig. 8a,b), where sparitic calcite filling in larger pores shows a complex internal fabric: larger irregular crystals 0.25 mm thick occupy the central part of voids, and diminish in size towards the walls of the aggregates. Sparite is sometimes juxtaposed with earlier phosphatic coatings. Other polyphase infillings contain scattered small pellets, indicating high-energy influx of solutions saturated with carbon dioxide. Precipitation of carbonates in Jamal did not produce either typical nodules in the groundmass or tubular forms of sparite in pores. Most

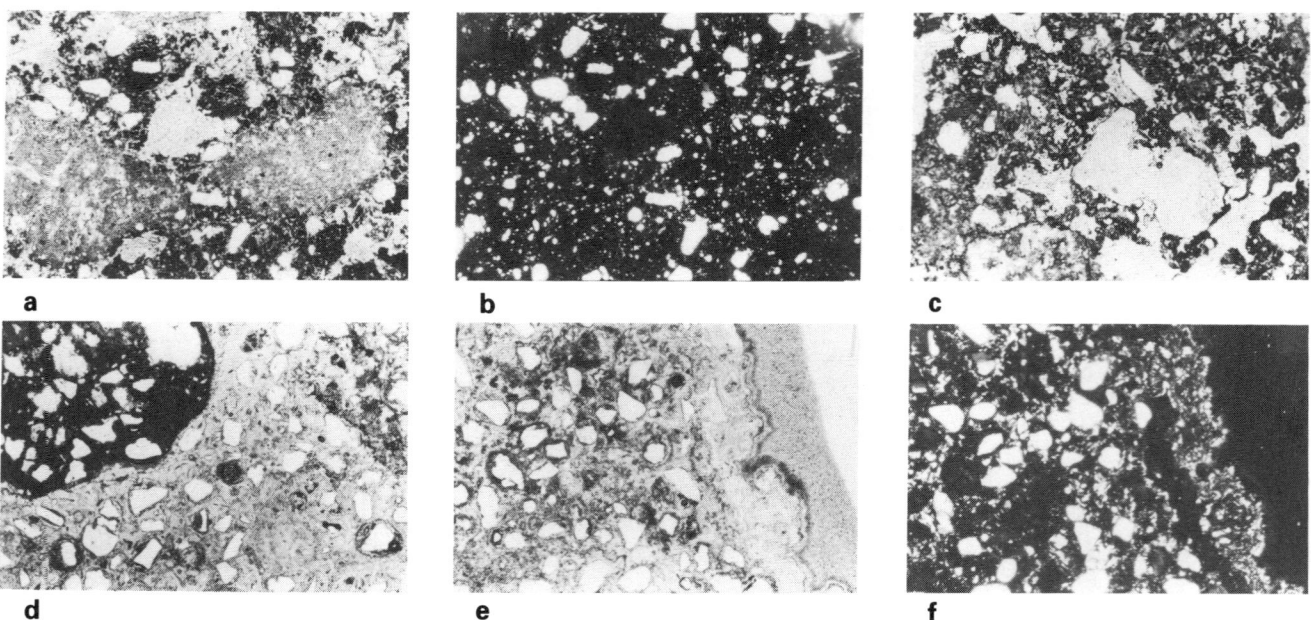

Figure 7.   Microfabric of secondary phosphatic deposits in Tabun and Jamal (the frame length is 2.8 mm). (a) Apatite pseudomorphs (light) with diffuse boundaries, partly assimilated by the groundmass, Tabun, layer (PPL). (b) The same view in XPL. Note abundant silty grains incorporated in pseudomorphs, testifying to high-intensity mixing of mineral and bone components by biological activity. (c) Bone pseudomorph (light; centre of the photo) with irregular outline embedded in the porous, aggregated groundmass, Unit 2a, Jamal Cave (PPL). (d) Part of the large phosphatic nodule with a complex internal microfabric testifying to several stages of formation, layer g, Tabun Cave (PPL). Note the dark-coloured core embedded in transluscent phosphatised matrix containing abundant mineral grains. (e) Transluscent mammilated hydroapatite coating in Layer E, Tabun Cave (PPL). (f) The same view taken in XPL.

probably the complex and contorted calcite infillings in Jamal are linked to episodes of inundation, the cause of which could have been either ground water rise or percolation of meteoric water through cavities in the roof of the cave.

d. Manganisation.

Manganese deposits occur in thin bands in the lowermost sediments of Tabun and more commonly in Jamal. Black exfoliated intercalations, 1 to 3 cm thick, were found at the transition between the disturbed sediments adhering to the wall (Unit 1) and the sediments of the central part of the cave (Unit a; Fig. 3). In thin sections, these intercalations show a clear ooid-type microaggregation, complete decalcification of the groundmass, opaque black flakes and coatings on the walls of fissures (Fig. 8c). These coatings were identified by X-ray diffraction as cryptomelan. Many of the black coatings, however, are anisotropic and red, suggesting association of manganese with iron, organic matter, clay and phosphorus. At high magnification by SEM (Fig. 5d) dark-coloured rounded 0.05 mm clayey pellets are covered with thin (several microns) coatings; microprobe analyses showed that the coatings are composed of Mn, mixed with clay and apatite (Fig. 6d).

The accumulation of manganese seems to take place separately from iron but in close association with phosphorus. The tiny pellets observed by SEM may be related to organic-rich materials, presumably of anthropogenic origin such as hearths. Organo-mineral complexes in these materials could have been destroyed by microorganisms under strong reducing conditions to release and precipitate Mn (McKenzie 1978). However, manganese mineralisation in Jamal could have also originated from redistribution of Mn-bearing minerals by ground water at the time of inundation and subsequent collapse.

e. Biological activity and humification.

In spite of strong weathering of mineral mass and organic remains, some biological features can be observed by microfabric analysis. For example, in dark-coloured layers in Tabun the microstructure is characterized by the presence of numerous channels, loose high-order aggregation and finely comminuted organic matter intimately mixed with clayey mass. In other layers pellets and rounded, partially welded bioaggregates can be discerned. These could have originated from the burrowing activity of soil fauna. Such activity also controls humification processes by comminuting and dessiminating organic materials and encourages their association with clay particles to produce organo-mineral bonds.

Processes of humification are also strongly influenced by soil microorganisms. Various signs of microbiological activities can be discerned. For example, transluscent phosphatic coatings, especially around and inside bone pseudomorphs or bioaggregates, seem to originate from the activity of particular algae species, as noted above. Different organic remains, especially chips of bones, also exhibit signs of microbiological transformations.

f. Anthropogenic features.

The most common sedimentological indicators of human occupation in prehistoric caves are lenticular ash layers and hearths. Their recognition in the field in Lower Palaeolithic sites is difficult and sometimes equivocal. With the help of micromorphology we could recognize throughout Tabun and Jamal sequences the presence of scattered pieces of charcoal, 0.5 - 0.9 mm across, which were secondarily disrupted and rounded by colluviation (Fig. 8d), as well as comminuted charred organic materials incorporated into microaggregates. Tiny fragments of burnt bone, still preserving its original texture, have been identified occasionally and may be related to a low-temperature fire. Alterations of ferruginised chips of bones in Tabun layer G (Fig. 8e,f) can also be attributed to heat. They show an intense red colour in plane light and 'fluidal'

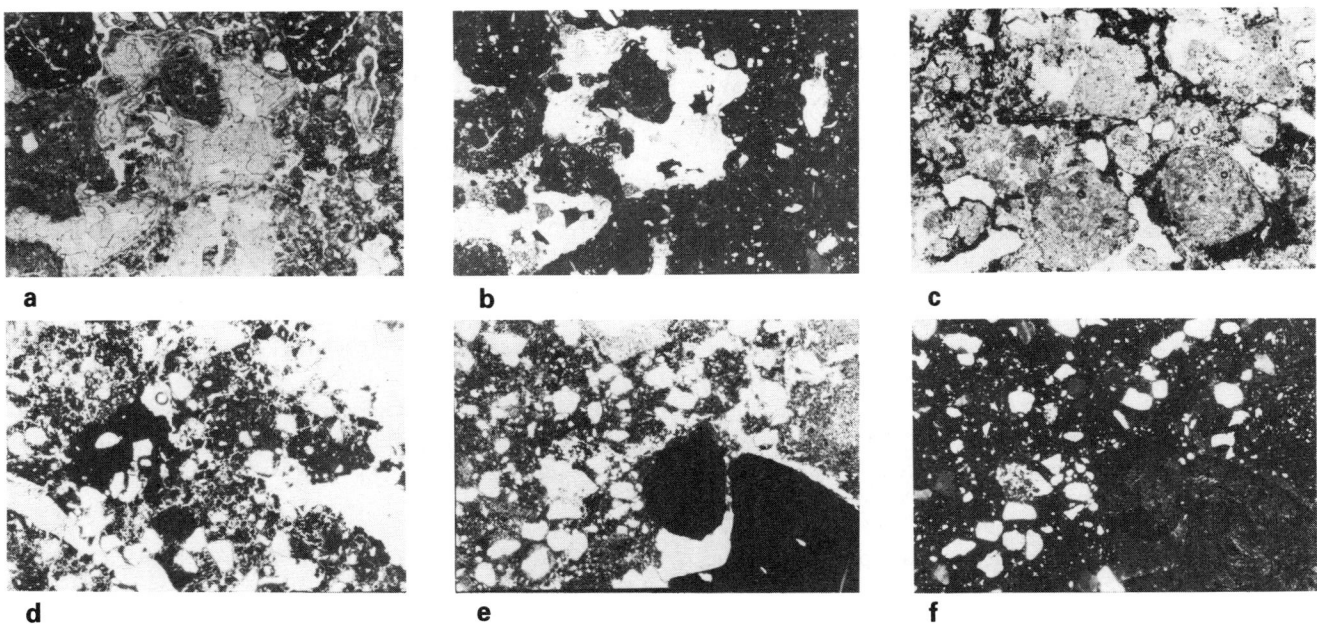

Figure 8. Microfabric of various chemical precipitates and anthropogenic features in Tabun and Jamal Caves (frame lenght is 2.8 mm). (a) Sparite void infilling containing rods of groundmass 0.2 - 0.05 mm in Unit 2, Jamal Cave (PPL). Note the irregularity of shapes and variability in size of calcite crystals. (b) The same view taken in XPL. (c) Jamal Cave Subunit 1a, a complex ooid microstructure with abundant black manganese coatings (PPL). (d) A charcoal fragment secondarily disrupted (black in the centre of the photo) incorporated in the groundmass with signs of pedogenic reworking, layer G, Tabun Cave (PPL). (e) A ferruginized chip of bone with broken edge (black in the right corner of the photo), presumably a result of heating, layer G, Tabun Cave (PPL). (f) The same view taken in XPL. Note the fluidal texture of bone pseudomorph contrasting the isotropic fabric of the matrix.

extinction under crossed nicols.

Dark-coloured lenses and layers, enriched with organic matter, may also be related to possible human occupation layers, in spite of the fact that in thin sections they resemble 'humified' horizons without clear evidence of human impact. The paucity of such features may indicate that anthropogenic materials in Tabun and Jamal have been strongly secondarily altered, or even erased, in the course of prolonged diagenesis. Furthermore, the manganese oxides, found in Unit a in Jamal (see above), could also be formed from charred organic materials by oxidation-reduction processes under gleying conditions.

g. Intensity of pedogenic reworking.

Laville *et al.* (1980) distinguished specific layers within cave sediments designated as 'cave soils'. These are morphologically similar to modern soils, *i.e.* showing A, B and C horizons. Such pedogenic profiles were not observed in Nahal Me'arot caves. However, on a microscale various materials are clearly organised in particular patterns, possibly by processes similar to pedogenesis. This arrangement of micromorphological features in various cave laminae is employed here to assess the nature and extent of pedogenic reworking.

Large variability of the groundmass fabric has been observed in different layers and units. Because of strong dissolution of bones and impregnation of clay groundmass by secondary phosphatic minerals, crystallitic fabric does not occur (Goldberg & Macphail 1990). This is because of the weak birefringence of apatite minerals and possible homogenisation by soil fauna. However, in layer G of Tabun, the clayey groundmass has a weak crystallitic b-fabric. In some patches, grains of coarse sand and larger microaggregates display a granular b-fabric, which indicates a certain mobility of thin particles and parallel orientation of clay. In general, their geometrical arrangement is similar to the red loam (Hamra) soils of the Israeli Coastal Plain (Wieder & Yaalon 1983). In Jamal the energy of

reorientation of particles within individual aggregates and ooids led to a stronger crystallitic b-fabric, despite the impregnation with phosphates.

Discussion and Conclusions

Systematic application of a polarising microscope and scanning electron microscope with microprobe to Tabun and Jamal cave sediments clarified the sedimentological and geochemical processes in the Middle Pleistocene layers of these prehistoric sites. Being close to the Mediterranean coast, the caves have been strongly affected by aeolian processes and by fluctuations of ground water, possibly related to sea level changes. In Tabun, most of the sediments are wind-blown sands and silts. The sandy nature is characteristic also of the Lower Palaeolithic sediments in the neighbouring Skhul cave (Garrod & Bate 1937; Ronen 1976), suggesting a similar source of wind-blown material. In Jamal the sediments are from more varied sources. In Tabun, a substantial increase of aeolian sand accumulation can be observed from layer F onward. From the middle part of layer E accumulation of detrital silt material prevailed. In general, during the Lower Palaeolithic, rates of accumulation were significantly greater than in later periods.

Our studies demonstrate that the sediments in Tabun and Jamal were strongly reworked by post-depositional processes related to anthropogenic and biological activity, chemical weathering, phosphatisation and later movement of soluble products, and ground water effects. Post-depositional transformations of various materials are thought to occur both at the stage of early diagenesis or much later, after inundation by ground water. Differentiation between pene-contemporaneous changes in archaeological layers in the caves and later ground water effects is still problematic (Weiner *et al.* 1993). In addition, we suggest that modifications of cave sediments can also be attributed to pedogenesis, which is expressed in microaggregation and homogenisation of fine-grained mass by burrowing animals, formation of clay fabric patterns, and migration and deposition of secondary constituents. It is worth stressing that in contrast to surface soils, abundant features in cave sediments can be related to anthropogenic activites: phosphatic pseudomorphs and nodules, charred organic materials, chips of probably heated bones.

It is interesting to note that some phosphatic nodules show a polyphase internal structure, and are composed of phosphorus, closely bound with silica, iron, titanium and manganese, in different proportions. We tentatively suggest that, following the dissolution of bones, phosphate reacts with various mineral elements (Fe-Ti phase, or iron, or manganese) which are released from silicates and oxides through natural weathering or anthropogenic impact. The complex morphology and composition of these features may imply that processes of weathering of bones and mineral mass could have initiated prior to inundation by groundwater. However, when ground water inundated the caves, processes of geochemical weathering were reinforced drastically. It was probably during this time that bones were almost completely dissolved in the Middle Pleistocene sediments in Tabun and Jamal. The sparite infillings superimposed upon phosphatic limpid coatings in voids testify to changes in water regime and pH of percolating solutions at this time.

The intensity of geochemical and pedogenic reworking within caves can vary in accordance with geomorphic situation, which controls the ability of fluids to migrate and to affect the original composition and morphology of particular components. For example, in Tabun, the sediments adhering to the wall, contain phosphatic and Fe-phosphatic pellets and nodules embedded in the calcite-rich media. More intensive transformations were identified around the karstic sinkhole, where the sediments are decalcified and contain indurated phosphatic nodules and crusts, which were cemented in the course of slumping. A particularly strong heterogeneity of sediments and degrees of weathering are exhibited in Jamal Cave. Here, secondary calcite, phosphatic and manganese deposits coexist within small patches, and bones have completely disintegrated. Ooid microstructure also testifies to the combined effect of repeated wetting and drying and short distance transportation of sediments after the archaeological layers were covered by later sediments.

In conclusion, we have demonstrated the usefulness of microscopic, submicroscopic and microchemical methods for the identification and understanding of sedimentological and geochemical processes in the formation of cave sequences. Various processes have been distinguished, namely accumulation of aeolian, colluvial and anthropogenic materials and their progressive alteration by both penecontemporaneous and later diagenetic processes.

Acknowledgements

We are grateful to Daniel Kaufman for his comments and assistance in editing the manuscript, Shimon Ilani and Michael Dvorachek for SEM facilities in the Geological Survey of Israel, and to Israel Hershkovitz and Irit Zohar for computer graphics. We would also like to thank the two anonymous reviewers for their constructive comments. The research was sponsored by the Israel Nature Reserves Authority, the Zinman Institute of Archaeology, and the Irene Levi-Sala CARE Archaeological Foundation. Special thanks are due to the Schussheim Foundation and to the Wolfson Family Charitable Trust for their generous support of this research.

References

BAR-YOSEF, O. & VANDERMEERSCH, B. (1981). Notes concerning the possible age of the Mousterian layers in Qafzeh Cave. In: Prehistoire du Levant. Editions du CNRS, N 598: 281-285.

BAR-YOSEF, O., VANDERMEERSCH, B., ARENSBURG, B., BELFER-COHEN, A., GOLDBERG, P., LAVILLE, H., MEIGNEN, L., RAK, Y., SPETH, J.D., TCHERNOV, E., TILLIER, A-M. & WEINER, S. (1992). The excavations in Kebara Cave. Current Anthropology, 33: 497-550.

BULLOCK, P., FEDOROFF, N., JONGERIUS, A., STOOPS,G. & TURSINA,T. (1985). Handbook for soil thin section description. Wolverhampton, Waine Research Publication, 152 pp.

COURTY, M.A., GOLDBERG, P. & MACPHAIL, R. (1989). Soils and Micromorphology in Archaeology. Cambridge University Press.

FARRAND, W.R. (1979). Chronology and Palaeoenvironment of Levantine prehistoric sites as seen from sediment studies. Journal of Archaeological Science, 6: 369-392.

FLICOTEAUX, R. & LUCAS, J. (1984). Weathering of phosphate minerals. In: Nriagu, J.O. and Moore, P.B. (ed.) Phosphate Minerals. pp. 292- 317.

GARROD, D.A.E. & BATE, D.M.A. (1937). The Stone Age of Mount Carmel. Vol 1, Oxford, Clarendon Press.

GOLDBERG, P. (1973). Sedimentology, Stratigraphy and Paleoclimatology of Et-Tabun Cave, Mt. Carmel, Israel. Unpublished Ph.D. Thesis, University of Michigan.

GOLDBERG, P. & MACPHAIL, R.I. (1990). Micromorphological evidence of Middle Pleistocene landscape and climatic changes from Southern England: Westbury-SUB-Mendip, Somerset and Boxgrove, W.Sussex. In: Soil micromorphology. L.A.Douglas (ed.), pp. 441-448, Elsevier, Amsterdam.

GOLDBERG, P. & NATHAN, Y. (1975). The phosphate mineralogy of et-Tabun Cave, Mount Carmel, Israel. Mineralogical Magazine, 40: 253-258.

GRÜN, R., STRINGER, C.B. & SCHWARCZ, H.P. (1991). ESR dating of teeth from Garrod's Tabun cave collection. Journal of Human Evolution 20: 231-248.

ILANI, S. (1989). Epigenetic metallic mineralization along tectonic elements in Israel. Israeli Geological Survey Report, GSI/12/8 (in Hebrew).

JELINEK, A.J., FARRAND, W.R., HAAS, G., HOROWITZ, A. & GOLDBERG, P. (1973). New excavations at the Tabun Cave, Mount Carmel, Israel. Paleorient 1: 151-183.

JELINEK, A.J. (1982). The Middle Palaeolithic in the Southern Levant with comments on the appearence of modern Homo sapiens. In: A.Ronen (ed.) Origin of modern man. Oxford, BAR International series, pp. 57- 104.

LAVILLE, H., RIGAUD, J.-P. & SACKETT, J. (1980). Rock

shelters of the Perigord. Geological stratigraphy and archaeological succession. London, Academic Press.

LUMLEY, H. de, FOURNIER, A., MISKOVSKY, J.-CL., BOUDIN, R.-CL., PENEAUD, P., BEINER, M., PARK, Y.-C., CAMARA, A., GELEIJNSE, V., SAAS, A., HOFFERT, M. & SCHAAF, O. (1981). Evolution geochimique du remplissage de la Caune de l'Arago a Tautavel, posterieure a la mise en place des sediments. Colloque International du Centre National de la Recherche Scientifique. Datations absolues et analyses isotopiques en Prehistoire. Methodes et limites. Datation du remplissage de la Caune de l'Arago a Tautavel, 22-28 juin 1981, pretirage, pp. 79-93.

McDERMOT, F., GRÜN, R., STRINGER, C.B. & HAWKESWORTH, C.J. (1993). Mass-spectrometric U-series dates for Israeli Neanderthal/early modern hominid sites. Nature, 363: 252-254.

McKENZIE, R.M. (1978). The manganese oxides in soils. In: Varentsov, I.M. (ed.), Geology and geochemistry of manganese, vol. 1, pp. 259-269, Hung. Acad. Science.

MERCIER, N. & VALLADAS, H. (1994). Thermoluminiscence dates for the Palaeolithic Levant. In: Bar-Yosef, O. and Kra, R. S.(eds). Late Quaterbary Chronology and Paleoclimates of the Eastern Mediterranean. Radiocarbon, Department of Geosciences, The University of Arizona, Tuscon, Arizona, pp. 13-20.

MÜCHER, H.J. & MOROZOVA, T.D. (1983). The application of soil micromorphology in Quaternary geology and geomorphology. In: Bullock, P. (ed.). Soil Micromorphology, 1. Techniques and Applications. Berkhamsted, A B Acad. Publ., pp. 151-194.

PENEAUD, P. (1978). La paragenese phosphatee de la grotte de la Caune de l'Arago, Pyrennes-Orientales. Thèse de Doctorat de 3 cycle, Universite de Paris, 162 pp.

PENEAUD, P. (1981). Evolution geochimique et paragenese phosphatee du remplissage de la caune de l'Arago a Tautavel. In: Colloque International du Centre National de la Recherche Scientifique. Datations absolues et analyses isotopiques en Prehistoire. Methodes et limites. Datation du remplissage de la Caune de l'Arago a Tautavel, 22-28 juin 1981, pretirage, pp. 95-100.

PORAT, N., SCHWARCZ, H. & RONEN, A. (nd, in press). ESR Dating of burned flint from the Hominid bearing Tabun Cave, Israel - the isochron method. Paper presented at the 60th Meeting of the Society for American Archaeology, Anaheim.

RONEN, A. (1976). The Skhul Burials: An Archaeological Review. Intern Cong. Preh. Protoh. Nice, Colloque XII: 27-40.

RONEN, A. (1991). Tabun Cave: the 1991 season. Mitekufat Haeven, 24: 149-151.

RONEN, A. (1993). Mount Carmel Caves - new discoveries in Tabun Cave. Nikrot Tsurim 19: 9-16 (in Hebrew).

SCHWARCZ, H.P., GRUN, R., VANDERMEERSCH, B., BAR-YOSEF, O., VALLADAS, H. & TCHERNOV, E. (1988). ESR dates for the hominid burial site of Qafzeh in Israel. Journal of Human Evolution 17: 733-737.

STRINGER, C.B., GRUN, R., SCHWARCZ, H.P. & GOLDBERG, P. (1989). ESR dates for the hominid burial site of Es Skhul in Israel. Nature, 338: 756-758.

TCHERNOV, E. (1981). The biostratigraphy of the Middle East. In: Prehistoire du Levant. Editions du CNRS, N 598: 67-97.

VALLADAS, H., JORON, J.L., VALLADAS, G., ARENSBURG, B., BAR-YOSEF, O., BELFER-COHEN, A., GOLDBERG, P., LAVILLE, H., MEIGNEN, L., RAK, Y., TCHERNOV, E., TILLIER, A.M. & VANDERMEERSCH B. (1987). Thermoluminiscence dates for the Neanderthal burial site at Kebara in Israel. Nature 330: 159-160.

VOGEL, J.C. & WATERBOLK, H.T. (1963). Groningen Radiocarbon dates IV. Radiocarbon, 5: 163-202.

WEINER, S. & BAR-YOSEF, O. (1990). States of preservation of bones from prehistoric sites in the Near East: a survey. Journal of Archaeological Science, 17: 187-196.

WEINER, S., GOLDBERG, P. & BAR-YOSEF, O. (1993). Bone preservation in Kebara Cave, Israel using On-site Fourier Transform Infrared Spectrometry. Journal of Archaeological Science, 20: 613-627.

WEINSTEIN-EVRON, M. & TSATSKIN, A. (nd1, in press). The Jamal Cave is not empty: Recent discoveries in the Mount Carmel Caves, Israel. Paleorient 20-22

WEINSTEIN-EVRON, M. & TSATSKIN, A. (nd2, in press). Palaeoenvironmental investigations in the Jamal Cave, Mount Carmel, Israel. Proceedings of the Symposium 'Nature and Culture', Liege, December, 1993.

WIEDER, M. & YAALON, D.H. (1983). Micromorphology of Hamra soil. In: D.Grossman (ed.). Between the Yarkon and the Ayalon. Bar-Ilan University, pp. 27-34 (in Hebrew).

QUATERNARY RESEARCH ASSOCIATION

The **Quaternary Research Association** is an organisation comprising archaeologists, botanists, civil engineers, geographers, geologists, soil scientists, zoologists and others interested in research into the problems of the Quaternary. The majority of members reside in Great Britain, but membership also extends to most European countries, North America, Africa and Australasia. Membership (currently *c.*1100) is open to all interested in the objectives of the **Association.** The annual subscription is £10.00 with reduced rates for students and unwaged members.

The main meetings of the **Association** are the Annual Field Meeting, usually lasting 3-4 days, in April, and a 1 or 2 day Discussion Meeting at the beginning of January. Additionally, there are Short Field Meetings in May and/ or September, while Short Study Courses on techniques used in Quaternary work are also occasionally held. The publications of the **Association** are the *Quaternary Newsletter* issued with the **Association's** *Circular* in February, June and October; the *Journal of Quaternary Science* published in association with Wiley, with four issues a year; the monograph series *Quaternary Proceedings* also publised in association with Wiley; the *Field Guides Series* and the *Technical Guides Series.*

The **Association** is run by an Executive Committee elected at an Annual General Meeting held during the April Field Meeting. The current officers of the **Association** are:

President:
Professor **F. Oldfield**, Department of Geography, University of Liverpool, Roxby Building, Liverpool L69 3BX, UK.

Vice-President:
Professor **J.J. Lowe,** Centre for Quaternary Research, Department of Geography, Royal Holloway, University of London, Egham, Surrey TW20 0EX, UK.

Secretary
Dr. **P. Coxon**, Department of Geography, Trinity College, Dublin 2, Ireland.

Assistant Secretary (Publications):
Dr. **W.A. Mitchell,** Faculty of Sciences, Luton University, Park Square, Luton LU1 3JU, UK.

Treasurer:
Dr. **J.E. Gordon,** Scottish Natural Heritage, 2 Anderson Place, Edinburgh EH6 5NP, UK.

Editor, Quaternary Newsletter:
Dr. **J.D. Scourse,** School of Oceanographic Sciences, University College of North Wales, Menai Bridge, Gwynedd LL59 5EY, UK.

Editor, Journal of Quaternary Science:
Professor **M.J.C. Walker**, Department of Geography, University of Wales, Lampeter, Dyfed SA48 7ED, UK.

Publicity Officer:
Dr. **D. Bridgland,** Department of Geography, University of Durham, Science Laboratories, South Road, Durham DH1 3LE, UK.

All questions regarding membership are dealt with by the Secretary, the **Association's** publications are sold by the Assistant Secretary (Publications) and the **Association's** subscription matters are dealt with by the Treasurer.

NOTE FOR PROSPECTIVE EDITORS

Quaternary Proceedings is an occasional publications series established to report the proceedings of important meetings held under the aegis of the *Quaternary Research Association* or organised jointly by the *QRA* and other scientific organisations. Frequency of publication depends upon the receipt of proposals of reports of meetings which provide the basis for a coherent proceedings volume. Proposals should be submitted in the first instance to:

Series Editor, Quaternary Proceedings:
Professor J.J. Lowe, Centre for Quaternary Research, Department of Geography, Royal Holloway, University of London, Egham, Surrey TW20 0EX, UK.

The following guidelines must be adhered to in the production of volumes for the *Quaternary Proceedings* series:

1. All issues must conform with previous issues of *Quaternary Proceedings* in terms of quality of materials, type-set and printing (of both cover and text pages), arrangement of headings, size of lettering, general design and the use of the *QRA* logo. Reference can be made to the most recent issue of *QP* for guidance, and further particulars are available from the **Series Editor**.

2. Scientific Conventions and preparation of text and figures follow the 'Instructions for Authors' for papers published by the *Journal of Quaternary Science*. These are summarised on the inside back cover of issues of that journal.

3. Contributions should be supplied both as hard copy and on diskette, using any of the common word processing packages.

4. All proceedings proposals must obtain the joint prior approval of the *PUBLICATIONS SUB-COMMITTEE* of the *Quaternary Research Association* and editorial staff of *John Wiley & Sons*. Mock cover designs and text pages should be submitted for approval along with other general information of publication proposals.

5. Prospective editors will be required to complete a PROPOSAL FORM and to follow the specified arrangements for type-setting and printing.

6. Royalties generated by sales of an individual issue are shared equally between the editors of the issue and the *QRA*.

7. Copyright for *Quaternary Proceedings* resides with the *QRA*.

8. An ISBN number exists for *QP4* (ISBN 0-471-95860-3) and all volumes in the series must conform with the specifications to which this refers. For advice on this matter, consult the publishers, *John Wiley & Sons*.